Eradicating Ecocide

Eradicating Ecocide
Laws and governance to prevent the destruction of our planet

POLLY HIGGINS

SHEPHEARD-WALWYN (PUBLISHERS) LTD

Other books by the author:
Earth is our Business – Changing the Rules of the Game
I Dare You To Be Great

First published in 2010 by
Shepheard-Walwyn (Publishers) Ltd
107 Parkway House, Sheen Lane
London SW14 8LS

Reprinted 2011, 2012, 2014
2nd edition 2015

British Library Cataloguing in Publication Data
A catalogue record of this book
is available from the British Library

ISBN-13: 978-0-85683-508-7

DEDICATION
To my darling husband Ian, my serenissimus,
you are the sunshine of my life

Contents

Note to the Second Edition

Now more than ever, the message of this book retains its relevance. Whilst some specific examples have been updated, the basic tenet – where political will fails, the rule of law can prevail – still stands. Since I originally wrote this book, a concept paper has been seeded into many governments across the world, two more books have been published, a global movement to End Ecocide has taken root, Earth lawyers (and non-lawyers) across the world are supporting Ecocide law. It continues to be my honour as the lead advocate for Ecocide law to take forward a law that puts in place one simple yet profound principle – 'first do no harm.'

One caveat – all mention of 'what we need to do'/'must do' by me has been eradicated. 5 years on and I now see the world rather differently. We have a choice – what I lay out here is simply an offering of a legal solution that you may (or may not) resonate with. Far from watering down my position, I believe that action undertaken of free will is manifestly more powerful than 'doing what we are told.' Ending the era of ecocide is our choice.

Polly Higgins
September 2015

Introduction

It was 9:56 p.m on the 20th April when the fire started on the Deepwater Horizon oil drilling rig. Pressure mounted inside the marine riser and as it came up it expanded rapidly and exploded. The fireball could be seen 35 miles away. 11 men were killed, 17 others were injured in the blowout. Workers had less than 5 minutes to escape. After burning for more than a day, the rig sank on April 22nd 2010. That afternoon a large oil slick was observed and it was confirmed that the wellhead, a mile below the water, was damaged and pumping oil into the Gulf. Conservative estimates stand at +800,000 barrels a day pouring into the Gulf of Mexico. It is an environmental disaster of epic proportions, one of the largest in U.S. history. BP, the third largest corporation in the world, saw 42% of their share price collapse within 48 hours.

When the oil reaches the coastline it will seep into the surrounding marshland, destroying not only the habitat of wildlife but also the very roots of the grasses that bind the land together and prevent it from subsiding into the Mississippi River delta and the Gulf of Mexico. The importance of the survival of the marshlands is not only vital to wildlife but also to humans; it is the physical barrier that lessens the intensity of fierce storms like Hurricane Katrina. When that disappears, a lot more goes than just the death of the local fishing industry and surrounding aquatic wildlife.

Stemming the flow defeated BP for eight weeks. The Macondo well was eventually capped on 15th July 2010 when BP engineers performed a 'static kill.' BP claimed they would conclude final 'bottom kill' operations in the second week of September 2010.

This occurrence demonstrates a consummate failure on many fronts. Firstly of technology, reliance upon which betrays the hubris of an industry that believes technology provides all the answers. Secondly, this has demonstrated a far larger abrogation of corporate responsibility not just to employees but to the wider ecological community. Thirdly, the failure to anticipate the potential failure of technology[1] led to the fourth failure, namely failing to have the means in place to stop the disaster. For BP and the other companies involved, their worst fear was the final litigation bill arising from the individual and collective damages and lost income from local fisheries. The enormity defeated our comprehension, not just in size of disaster, but also in terms of identifying effective legislative controls.

It could have happened to any number of oil companies. The Deepwater Horizon tragedy is a warning of what can come in the future; for all the confidence of relying on technological innovation it just takes one small failure to result in catastrophe.

At the moment, aside from potential litigation for damages and loss to individuals there is no proper legislative mechanism that addresses the vast environmental impact of the oil spill. Quite simply, the laws and governance required to prevent such a disaster from happening again do not exist. The law as it currently stands, is not fit for purpose. Instead BP have the 'right to kill' the ocean without consequence.

Law can be employed creatively and constructively, effecting overnight change. Innovation can be nudged to suddenly flow in a very different direction, sometimes in unexpected areas. Law can close one door and open another. Law can inadvertently create a positive discrimination. Law can change our values and understanding. The inverse can also be true. Sometimes laws are put in place that directly or inadvertently cause damage and destruction. These are the laws to be rooted out and transformed. Change is what is required here, nothing less, if we are to change the current course of our trajectory of destruction. Sir David King, ex-chief scientific advisor to the former UK government, warns of a century of resource wars,[2] an era of ecocide. By asset-raiding our natural capital we deplete our resources to such an extent that conflict and war over the remaining few spoils is inevitable. It is a certain and rapid escalation into anarchy, death and destruction of epic proportions.

Eradicating ecocide requires radical and bold decisions. To eradicate ecocide, the means of doing so are embedded in the words themselves. We literally have to derail this unstoppable train of destruction that we have created. Applying the brakes gently is not going to work; it is a juggernaut that has acquired such powerful momentum that it is careering out of control. To stop it takes bravery, from those on the outside pulling up the railway track, and from those on the inside pulling the emergency cord. If both are done skilfully and quickly, very few will be hurt and the train will come safely to a sudden halt.

To eradicate means to pull up by the roots, to eliminate at source, overhaul. The Late Middle English origins of the word are derived from the Latin *radix*- 'root', *eradicat*- 'torn up by the roots'. We can pull up those roots, those railtracks. The word ecocide is prefixed with 'eco'; it derives from the 16th century Greek word *oikos* meaning 'house, dwelling place, habitation, family.' The suffix 'cide' means 'killer', from the use of the French -*cide*, from Latin *caedere* 'to strike down, chop, beat, hew, fell, slay.' To eradicate ecocide means to forcibly remove the systems that are killing and destroying our habitat.

Ecocide is like the virulent Japanese Knotweed – it spreads out of control, sucking the life out of all that comes in its way, strangling the life out of the very air we breathe. To stop it, it has to be eliminated literally at the roots. In oil industry terms, tackling the problem at the root is referred to as turning off the 'upstream', closing off the source. This means stopping the processes that extract and deplete the natural capital in its raw state. Do that and the downstream operations that are dependent on the life-force of the upstream operations shudder to a halt. In oil terms that includes the oil refiners, the product distribution terminals, the trucks, pipelines, marine transportation, the retail gasoline marketers. The wholesale, industrial and commercial customers no longer have their commodity. Sounds radical, but all those downstream operators will be required for mobilizing mass innovation in the opposite direction. The constructive use of those skills can be applied elsewhere, and very fast.

Today we are faced with a collapsing habitat – threats of increasing instability of climate change and the breakdown of national economies inevitably lead to the conclusion that governance of our planet has failed. As we destroy this planet so our eyes are now set on the potential to pillage from other planets, governments racing to see who can be the first to extract uranium from Mars, water from the Moon. Somewhere along the way, we have lost sight of our role as stewards. Until we know how to clean up the mess we have created in our own house, surely it is presumptuous to believe we can merely take from another.

It is no coincidence that the word economy has its roots in the Greek word *oikonomos,* which when taken in its constituent parts of *oikos* and *nomos* provides us with the etymology of the word. *Nomos* means 'manager, steward;' *oikonomos,* or economy, is stewardship of our dwelling place. Our largest habitat is, of course, the planet itself. Where our habitat is damaged and destroyed, humans suffer. Critics of environmentalists who claim we are failing to consider human needs are missing the point. Without the wellbeing of the ecology of our planet, our wellbeing suffers.

On a smaller scale, each of us mirrors our global habitat. Our bodies are our most immediate dwellings, composed of over 70% water, and so too is planet Earth covered by over two-thirds water. As we contaminate our own internal systems we start to suffer. Outbreaks, rashes, fevers are all indications of the body trying to rebalance and purge the poison. More serious and long-term abuse and exposure to toxins can lead to cancer, long-term disease. An acidic environment leads to loss of health and can result in pollution of our blood cells, the very life force our body depends on. This in turn opens us up to fatigue, infection and a downward spiral of illness with all its attendant symptoms if matters are left unchecked. Attending to our health and the health of others, not just human but of all life, is crucial if we are to optimize our wellbeing. Therefore, as we do locally, so too we can do globally for the larger habitat we inhabit. For example, pollute the waterways by dumping toxic waste and the result is acidification; marine life is compromised and/or dies, the tributaries carry the contaminated water that seeps into nearby land poisoning the foodstuffs grown therein. We eat the fish and food, we drink the water, taking it all into our own habitats knowingly or unknowingly. The damage is pervasive: what is damaged here, comes out over there and remains in the system. Everything has consequences, and destructive practices come back to affect us and cause harm, if not locally then elsewhere and to others. Pollution created in one place can affect another quite randomly – Inuit women of Alaska have high levels of DDT in their breast milk but until notified, they had no knowledge of it.[3] DDT was banned 15 years ago, yet its impact on future generations is playing out today in many ways. Pollution, transboundary in nature, does not adhere to man-made borders.

Nothing less than a radical and rapid shift from protection of private interests to protection of public interests is what I call for. We can rebalance the scales of justice by implementing laws to ensure the wider community is provided for, not just humans but also other species who are (or are at risk of being) adversely affected in a given territory. To remain rooted in protecting the private is to set apart, whereas place the community at the centre and we ensure the well-being of our global commons.

Teetering on the Brink

As a global society we stand at the precipice, teetering on the brink of shifting from independence to interdependence. Just as that moment comes when a teenager becomes an adult and starts to take responsibility for the consequences of his or her actions, so now is civilization as a whole poised to step into that space. Sometimes the key tipping points are identifiable – it can be triggered by an event, or series of events; often it is not. Just as with the teenager who seemingly overnight has become an adult, so there is that recognition that the shift in consciousness has somehow taken place. The teenager has become an adult, no longer striving for independence but now embracing interdependence, creating relationships, understanding the intrinsic value of the self and others.

A shift in consciousness is not a gradual process. Rather, it is akin to a build up of pressure, sometimes resulting in outbursts, tantrums, the attempt to cling onto practices that are of no long-term benefit, but guarantee short-term satisfaction. So it is with society, with our banking systems, our blind use of damaging extraction of fossil fuel, unaccountable practices, mistrust and short-term trading gains at the expense of people and planet. Our governments speak of the necessity of 'independence' and 'security' of supplies. Yet true resilience, as seen in nature, is found in forging interdependent relationships based on sharing and the creation of abundance. Security concerns will become obsolete. Very soon a tipping point will be reached. We do nothing and all systems, ecological and human-made, will implode and we become mere spectators of our own destruction. Or, we collectively step forward to take responsibility for our actions and put in place the means to ensure the well-being of our *oikos*. Once we have that shift in consciousness where we assume collective responsibility, that will be the true journey of social progress.

In legal terms this means the journey of moving away from laws premised, either deliberately or accidentally, on compromise. Laws that are often put in place with the promise of robust sounding enforcement restrictions. Such laws are often no more than a sop to appease public opinion or are weak instruments which have been watered down by intense corporate lobbying. They can have the appearance of being radical, or are touted as innovative, but the reality is that they ensure the status quo remains. These types of laws are dismissed by lawyers as 'compromise laws.'

These are often laws that have been put in place to minimize – but not prohibit – an activity. Instead, the problem continues, often in time worsening. History demonstrates that laws dealing directly or indirectly with the environment, for example pollution reduction measures, have comprehensively failed where reliance is placed on incremental mechanisms, limitations, efficiency measures and permit allocations. In reality corporations simply deal with these measures by making any breach, and the consequence thereof, a part of the profit and loss account of its operations. Modern-day climate negotiations are a further example; it has spawned a rash of laws premised on compromise which serve only to advance the interests of industry, not people and planet.

The very idea of negotiating our way out of climate change is a premise that we have accepted without question, despite the impossibility of succeeding. The Kyoto Protocol is a commodity document, a trading document; it has nothing to do with protecting people and planet and is quite simply not fit for purpose. There are missing rules, ones that create barriers to halt the gambling of our planet. As a lawyer who has experience of protracted divorce proceedings where two people cannot agree on the division of goods, it occurs to me that negotiating between 194 countries, with all their vested interests and teams of negotiators to agree on a range of topics, is a recipe for years of fatal compromise and disagreement.

Radical law, on the other hand is law that completely changes the landscape. Sometimes it is a law in a totally different arena that can do this. For instance, the first British Canal Enabling Act in 1759 opened the door to cheap commercial transport. Another, this time in the States, was the 1862 American Revenue Act, which placed an excise tax on alcohol. Both of these pieces of legislation, innocuous enough in themselves, removed a block and opened the floodgates respectively in each country (and subsequently throughout the world) to rapid uptake of coal and oil. Such laws are acupressure points, the levers which facilitate the smooth flow of energy (in this case literally as well as metaphorically) in a completely different direction for innovation, investment and policy. At other times it is new law (and new language) that has had to be invented to prohibit destructive practices. The international law of Genocide is the most telling example of the 20th century.

JUST CALL ME TRIM TAB

Walking through Lincoln Inns Fields behind London's Royal Courts of Justice on a spring afternoon with Victoria, a wonderful and wise friend of mine, I was discussing application of a law of Ecocide. "What we need are trim tabs", she said. Not knowing what this was, she explained to me that a trim tab button is to be found on cruise liners, to be pressed when the ship is to turn in the opposite direction very fast. While everyone else is flailing around in the water expending enormous amounts of energy trying to push the ship around manually, in the cruise liner the only energy that is expended is that which it takes for one person to press the button. The cruise liner turns without any further pressure. The law of Ecocide, Victoria explained, is a trim tab.

Buckminster Fuller, (1895–1983) a self-described 'compre-hensivist' who took a long view of history, had this to say when interviewed by *Playboy* in 1972:

> *Something hit me very hard once, thinking about what one little man could do. Think of the Queen Mary – the whole ship goes by and then comes the rudder. And there's a tiny thing at the edge of the rudder called a trim tab. It's a miniature rudder. Just moving the little trim tab builds a low pressure that pulls the rudder around. Takes almost no effort at all.*

Fuller had been a naval officer in World War I and so experienced the use of a trim tab first hand. The slightest pressure on the trim tab moves the rudder, which in turn directs the ship. He applied the metaphor to our lives. We are all trim tabs, tiny pivots affecting the overall direction of humanity. As Fuller advised, it is time to take a long view. Zoom out, look at where we've been and where we might be going. Fuller urged leaders of government and industry to focus not on weaponry, but on what he called 'livingry' – the tools necessary to promote peace and prosperity for the entire planet's population. Cooperation, not competition, would signify the next step of human evolution. The law of Ecocide is a trim tab to take us to 'livingry.' The epitaph on Bucky's gravestone was fitting: just '*call me trim tab.*'

HUMAN RESPONSIBILITIES

Over time as our understanding expands of the impact our activities have, so too comes the recognition of an expanding responsibility. The shift in consciousness required here is an espousal of our collective responsibility for the ecology of our planet.

We are only one of many millions of species, but our role as humans is particular. Whilst we are not the world's largest species in terms of numbers, our impact on our environment has been by far the most influential. We have built a world of beauty, but in the process destroyed far more in pursuit of our perceived path to progress. For life to reintegrate rather than disintegrate requires us to redefine our role.

When we step into the space that defines us as fully matured human beings, and fully embrace the meaning of being a *homo sapiens* (Latin: wise or knowing human), we shoulder with ease our individual and collective responsibilities. Our role shifts from conqueror to provider, from corporation to co-operation, from owner to steward. Competition, all too often bolstered by scarcity, blame and fear, is replaced by collaboration fostered by trust and, most importantly, love.

We can remedy all of this. We are learning to apply different language and different laws. Time has demonstrated that some laws work, some laws do not. Flip our normative from property laws to the use of existing trusteeship laws, and we accept our duties, responsibilities, obligations. For those of us who live in a world of governance by written law, bridges are required to take us to this new world. Law has an important role in providing some of the crossings to get us there. Some new laws, such as the law of Ecocide, will provide the footpaths in our new responsible world. Some existing mechanisms that have been lying defunct such as the United Nations Trusteeship Council can be dusted down, taken out of abeyance and put to good use once again. We can do all of this and so much more.

The implications of the proposals set out here for society, the environment and climate change strategy are enormous, both at international and national level. By applying both new and existing laws, we can impose upon international corporate activity, banking and governments a global standard of care, a pre-emptive obligation of ecological responsibility and accountability for and of our natural world. International regulatory frameworks can be simplified, finance can be mobilized and infrastructure strategies can be exponentially accelerated. In so doing, a rapid transition to a cleaner world can be assured, without undue reliance on carbon markets, failed voluntary (and non-voluntary but unenforceable) mechanisms and compromise legislation.

At time of writing the first edition of this book, the world grappled with one of the largest manmade ecological disasters known to man. The tragic Deepwater Horizon Gulf oil disaster has served to demonstrate a shortcoming that all – banks, governments as well as corporations and society in general – underestimated. Existing laws are deficient. This singular event challenged our basic and fundamental assumptions on what is now deemed to be acceptable destruction of our ecology.

This book is applying Bucky's advice; it zooms out, to explain where we've been in our application of law to prevent ecocide in its various manifestations and where we might be going. It zooms back in to examine set examples and to set out guidance and legislative recommendations for use with immediate effect. All that is set out here is applicable to all nations and to all peoples, not only at an international level but also regional, national and local. This is a book not only to read but to be put to good use; to be used by decision-makers, policy-makers, law-makers and co-creators of the new world. Whether you are a member state representative in the United Nations, an activist, a judge in an international court, a community spokesperson, a congressman sitting on an enquiry into an oil spill, a lawyer, an MP, a town councillor, a concerned parent – all of us have our role to play. Trim tabs each and every one of us. In short, anyone who has a say in decision-making processes that have an impact on the community, be it with regard to one's own local community or the global community. In microcosm and macrocosm, the principles are universal and applicable to us all.

Polly Higgins
London
30th July 2010

The International Year of Biodiversity
The International Decade for the Culture of Peace and
Non-Violence for the Children of the World 2000–2010

Part 1

HOW LAW CAUSED THE COMMERCIAL TAKEOVER OF THE WORLD

'Nobody made a greater mistake than he who did nothing because he could do only a little.'

EDMUND BURKE, Irish statesman, author, orator, British politician, and philosopher (1729–1797)

Chapter 1

TAKING STOCK

S OMEWHERE along the way, like a train that meets a forking of tracks, the decision is made to take one route over another. A simple shift in the direction of the track points and the train will take the left fork resulting in a destination many miles away from the other. One little decision can lead to a completely different route being taken – and a different end-point. The decision may seem insignificant at the time, but the small and almost imperceptible shift brings with it enormous ramifications. One train, two divergent routes. Two different outcomes.

One Planet, Two Divergent Approaches.

View the planet as an inert thing and a monetary value can be imposed. In one fell swoop the planet becomes a commodity. The planet becomes property to trade, an asset upon which a price has been imposed. As a consequence, (wo)man's dominion over land, his/her right to extract and diminish the capital of the planet as he/she wishes, guarantees the asphyxiation of life.

This approach has informed our environmental legislation almost exclusively[1] since the 1970s. Permit allocation, soft industry requirements and inadequate enforcement provisions are the modus operandi. Businesses that damage the planet continue apace, hand-in-hand with ecological devastation – at a price. In real terms the price is far higher than the nominal pecuniary fines levied against those caught exceeding their allocated limits. By reducing the planet to a commodity and ensuring industry is legislatively cushioned, damaging business practice is legally protected. The value of life is of no consequence but value of profit is. In the UK, as in many other countries, the duty to act in the interests of profit above other interests is enshrined in law. Section 172 of the Companies Act 2006 is the legal duty to promote the success of the company. When exercising this duty the director is required to 'have regard' to various non-exhaustive list of 'factors' listed in s.172. To 'have regard' is a far lower legal standard to attain than to 'prioritize.' The difference is determinative: the former can be given the most cursory of considerations, whilst the latter demands a determination according to the importance of the factor. 'Success' is not defined in the Act. The Department of Trade and Industry's guidance to the Bill suggests that a success in relation to a commercial company is considered to be its "long-term increases in value". It is suggested by the DTI that a director will exercise the same level of care, skill and diligence as he carries out any other functions in deciding which 'factors' he will take into consideration when making a decision subject to his overall responsibility to the success of the company. The implication is that where a factor is deemed a conflict of interest, then it can be disregarded.

Current climate negotiations reinforce customary rights for business – the right to emit, the right to be inequitable, the freedom to destroy the planet. These are the cushions upon which industry sits; soft governance ensures profits remain secure. The Kyoto Protocol is a document that actively facilitates trading on these terms (permits to pollute, carbon trading mechanisms). Thus an international business has been created to address one symptom (the escalation of greenhouse gases) rather than the problem (the damage we are wreaking on the planet). Twelve years on from its instigation and we know that both the mechanism applied and the ascribed rights have comprehensively failed to stop damaging practices. Instead, we have enslaved the planet for our own ends.

THE INDUSTRIAL REVOLUTION
Treating the planet as a commodity became firmly established as the norm in the Western World with the advent of the Industrial

Revolution. Rapid expansion of industry spread throughout the UK into Western Europe and North America, then onto much of the rest of the world. Agricultural based manual and animal labour shifted towards large-scale use of steam-powered machines, fuelled by abundant and cheap coal. The steam engine took over from the water wheel thereby enabling the powering of a wide range of manufacturing machinery. Commercial activity was no longer restricted to locations where water wheels or windmills could be used. Coal burning to heat water and create steam opened the door to new opportunities; coal could be transported and crucially could be used for transport. Burning coal created energy to power ships and railway locomotives. Increased mobility increased speed of transaction and trade activity. There was an explosion of activity; textile industries were mechanised and iron-making techniques developed exponentially assisted by the introduction of canals, improved roads and railways. Coal was the common denominator; it liberated the expansion of trade and transport.

'Canal mania' had grabbed the imagination of the nation in the mid-18th century. Transport had until now been a costly affair, not to mention dangerous. The upkeep of roads, horses, carriages and men as well as the turnpike toll payments made movement of heavy goods cumbersome and expensive. Proving to be an enormously effective means of transport, an innovative Duke decided to set up his own canal to link his mine to Manchester some 42 miles away. Suddenly every entrepreneur saw the potential for transforming the movement of goods.

It took very little to convince Parliament. Government stepped in and provided the necessary political support. Canal Acts and Navigation Acts were hastily written and implemented as each new canal was conceived. Outside of coal mining, it was the largest job-creation scheme of its time. Thousands of workers were employed, larger and greater canal trade routes were planned across the country, links were made to existing city rivers. Within an incredibly short period the geographical and economical landscape of Britain had changed dramatically. Brewers were no longer confined to selling their ales nearby: a horse drawn barge loaded with barrels could be delivered into the heart of London within two days. The fragile porcelains of Joshia Wedgewood, free from the risk of being shattered as wheels clattered over unpaved rough roads, could now be floated serenely and safely downstream to arrive in one piece in the shops of all the major cities.

The public directly benefitted too. The reduction in transport costs resulted in the price of coal dropping dramatically which in turn facilitated the switch from wood as the predominant fuel (the forests of Britain were by now vastly depleted). Coal was the number one cargo for canals. The savings were astounding: one horse could haul up to 400 times as much as a single pack-horse. [2] Coal was now readily accessible, readily available and readily used. Parliament took a pro-active policy of implementing enabling laws for all canals requested. The Canal Enabling Acts provided the necessary legislative leverage required to ensure coal became the new fuel of choice. The first of many enabling Acts was passed in 1759, triggering a decade of enormous activity. It was the trim tab that immediately unlocked the availability of coal at vastly reduced price.

The dramatic upsurge of the use of coal, however, came at a price in other ways; increase in burning brought increase in pollution. Akin to a disease, the spread of coal pollution was detrimental and degenerative, both directly and indirectly affecting humans. This was no benign cancer; this was the beginning of over 200 years of large-scale damage, destruction and loss to the environment, contaminating the air, land and waterways.

The origin of this particular cancer was London. By the late 18th century no other city in the world consumed such enormous quantities of fossil fuel, unwittingly exposing the inhabitants to such high levels of pollution. In the year 1800 one million tons of coal were burned to satisfy the rapacious demands of a million Londoners, throughout the UK the number reached 15 million tons. [3] The burning of coal unleashed a wave of pollution that was to engulf the world, fuelled by Britain's expansion as the most powerful manufacturing, trading and imperial power that the world had ever seen.

By the late 19th century technological development brought the internal combustion engine and electrical power generation. The impact of the use of coal on society was enormous. Within 100 years society's relationship with the planet had been changed irrevocably – certainly for those who were the pioneers of the industrial world. The emergence of factories and consumption of immense quantities of coal and other fossil fuels gave rise to unprecedented air pollution. The masses were mobilized; corralled into factories that were built near to coal mines creating urban centres and pollution hotspots. It was not only the well-to-do who could travel with greater speed and in greater numbers, and increased production was facilitated by the greater ease and speed of transporting merchandise. The availability of coal-fuelled heat and gas made from coal for indoor and outdoor lighting irrevocably altered the industrialized man's lifestyle. Profits were to be made directly or indirectly out of coal with little thought or concern for the health and wellbeing of the planet, nor initially for the people who were adversely affected by the damage and destruction wrought in the wake of industrialization.

Common understanding of pollution in 19th century initially did not make the causal link between fossil fuelled industrial activity and damage. Although smoke from the combustion of coal was visible to the eye, and particulates left their pernicious residue on skin, clothes and in the air, few viewed it as detrimental to either human health or the environment. Pollution was deemed to be something altogether far more sinister; it made you ill and was contagious. Noxious gases given off by decaying plant and animal matter were the danger (miasma, or 'bad air') and smoke from coal burning was viewed as an antidote. According to prevailing thought at the time, shaped by the fear of cholera and the Black Death, acids and carbon in smoke were powerful disinfectants. The pong of putrefaction was evidence enough of pollution. Pollution was airborne and smoke, it was believed, provided a blanket service in soaking up the offending odours.

The widespread view industrialized man and woman held, of nature as harmful, had successfully, if erroneously, long been germinating among city dwellers; putrefaction and decomposition of animal and plant matter had brought infection and disease for centuries. Nature was thus to blame. Cesspools, sewers, marshes and canals – anywhere that decomposition of biomass was to be found was feared. For centuries people had been complaining that London's rivers were smelly and polluted. Tanneries discharged their toxic cocktail of fermented offal, skin scraps and dog faeces into the river, accompanied by rotting animal parts discarded by butchers, human sewage and other industry by-products. Sal ammoniac (ammonium chloride) was one of the first industrial applications of a chemical; it was used extensively in tanning and dying processes, also to dissolve metals and later, applied in fertilizers. [4]

In cities, viewing nature as detrimental and something to be controlled had replaced the view of nature as beneficial. What was not understood was that it was humans' own desecration and inconsiderate disposal of nature that created the pollution. In addition to the polluted waterways, the burning of our natural resources created more of that which we feared the most: disease and ill-health.

The Discovery of Germs

On 31st August 1854, in Soho, a major outbreak of cholera struck. It was not an isolated case, but it was the worst yet. Within three days 127 people on or near Broad Street died. In the next week, three quarters of the residents had fled the area. By 10th September, 500 people had died and the mortality rate was 12.8 percent in some parts of the city. By the end of the outbreak 616 people were dead.

A recent increase in migration of people into London in search of work triggered a serious problem with the disposal of human waste and a lack of proper sanitary services. Disease spread quickly through large conurbations around workhouses and factories; settlements that were cramped with no organized sanitation. Many basements had cesspools of nightsoil underneath their floorboards. The cesspools had reached capacity and were overrunning; the government responded by dumping the effluence in the River Thames.

The physician John Snow was a sceptic of the miasma theory that declared 'bad air' as the cause of cholera and the Black Death. He decided to investigate. He identified the source of the outbreak as the public water pump on Broad Street (now Broadwick Street), and subsequent examination of the water and pattern of contamination led him to believe that it was not due to breathing foul air but from drinking the water. It was discovered later that this public well had been dug only three feet from an old cesspit that had begun to leak fecal bacteria. The pump handle was removed. By that time Snow suspected the contamination had already receded. Nevertheless, Snow's study proved to be a major event in the history of public health. His findings resulted in the replacement of the miasma theory with the germ theory.

Smog continued to fill not just the cities but also the lungs. By mid-19th century, and for the first time in world history, more people in Britain lived in cities than in the countryside. Coal burning continued to escalate exponentially both in the home and in the workplace. Soon doctors and physicians found their surgeries filled with patients suffering from respiratory diseases. Scientific breakthroughs and understanding of bacteria began to shape public opinion. Burning of coal released not only toxic ash, smoke and soot but also a whole range of impurities including sulphur dioxide, volatile hydrocarbons and carbon dioxide. Smoke, it would seem, could no longer be justified as beneficial. On the contrary, its detrimental impact was only beginning to be grasped. But the damage had already been done. The pollution train had metaphorically left the station to much fanfare and celebration, and with it the toxic plumes of smoke billowed their way down the line of time.

The train is one of the great symbols of progress of the industrial era, and indeed it was. Great railways were built, and stations to match: grand and beautiful edifices celebrating the arrival of a new era in the major metropoli. They were the palpable expression of the success of technology, design and man's dominion over his land. The 1830s were the golden era of travel: George Stephenson (1781–1848) was the genius behind the construction of the Liverpool to Manchester railway, the first to be floated across a vast tract of peat bog, and his locomotive, the Rocket. Both the train and the railway track were engineering triumphs that spawned thousands of miles of railway construction across the UK and the world.

Trains fuelled by the very coal they were transporting brought the black gold to the centres of commerce, stoking the fires of industry which in turn discharged their industrial chemicals into the already burdened and polluted waterways, poisoning the nearby lands. Cities and urban centres of industrial activities became increasing malodorous and noxious. People, property and land visibly deteriorated. The connection between industrial activity and pollution finally began to dawn, as the smog filled the air and the acid rain poured down. Blocking sunlight and nasal passages of politicians and people alike, from the mid-1840s several attempts were made to introduce laws to require owners of furnaces to reduce smoke emissions. However the recently acquired power and wealth of the new industrialists, who effectively lobbied Parliament, was not to be easily relinquished. The first pollution control Act, the Smoke Nuisance Abatement (Metropolis) Act, was only passed in 1853, but its effect was limited.

Parliament's response to pollution in general was tardy, cautious and piecemeal but one piece of legislation proved pivotal. In 1863 the Alkali Act was passed. Initially it addressed just one substance; gaseous hydrochloric acid from the Le Blanc alkali works (whose soda served the soap industry). At the time, severe problems were experienced near industrial plants manufacturing alkalis such as sodium carbonate and sodium hydroxide: emissions included hydrogen chloride, which was converted into hydrochloric acid in the atmosphere, causing extensive damage to vegetation. The Alkali Act required that 95% of the emissions should be arrested, and the remainder diluted. This had a dramatic effect: prior to the Act, emissions from alkali works were almost 14,000 tons annually, but after it came into force this was rapidly reduced to only 45 tons. Manufacturers were required to condense their gas and a team of inspectors were put in place to ensure they did.

In 1874, under a second Alkali Act, an Inspectorate was created to police all heavy chemical industry that emitted smoke, grit, dust and fumes. Although the extent of application widened, responsibility was watered down. Instead of applying strict and near 100% emission prevention (which had proven enormously successful in regulating the soda industry), industrialists were simply to apply '*the best practicable means*' to tackle pollution problems. This Act provided the foundation of air pollution policy in the UK for the next century. The Inspectorate was not very effective throughout most of the 20th century: between 1920 and 1967 there were only three prosecutions under the Act. Even in the early 1970s, a time of much increased environmental awareness, there were only 20. The message sent out to industry was one of leniency and complicity in their polluting activities. Industry requirements to use the '*best practicable*' environmental option, or the '*best available techniques not entailing excessive cost*' to control emissions led to a liberal interpretation of the law. Together with amendments, the Alkali Act became the first large-scale legislative control of industrial pollution in the UK. Its journey from near zero-tolerance application to loosely defined requirements fostered an atmosphere of acceptance of protecting industry's interests, with soft policing becoming the accepted norm. [5]

The journey of the Alkali Act demonstrates two crucial aspects that apply to all legislation governing commercial activity. Firstly, the intent of the legislation was to ensure complete eradication of a problem. The outcome desired was the elimination of the destructive practice. Industry could either implement a solution or stop using alkalis. In this case, it applied a solution to the problem; it passed the acid vapours through water.

Secondly, the subsequent Alkali Act of 1874 demonstrated a very different approach. The intent to eliminate pollution was replaced with a compromise. Within a decade the legislative language of prohibition had been replaced with the reasonability test. '*Reasonably practicable*' provisions are relative at best and never absolute, advocating efficient use rather than removal of the problem. Instead, cost becomes the determining factor in deciding the outcome. If industry determines that a risk is too far removed from the cost of preventing it happening, the company has a valid defence for not putting in place any further safeguards. In other words, a company can argue it was not reasonably practicable to pay out additional money to ensure health and safety concerns were addressed, as at the time of making the decision they had assessed the risk of damage as very small (although subsequent events may, and often does, prove the initial analysis to be wrong). Prohibition of a damaging or destructive practice was instead replaced with unspecified conditions of trade. What constitutes '*reasonably practicable*' provisions is largely determined by industry built on existing practice, which in time improves slowly through natural evolution rather than necessity. Prohibition laws, in contrast, impose necessity and industry reinvents its wheels speedily in response. Prohibition of a damaging practice, as demonstrated by the first Alkali Act, was an absolute protection of society's and environmental interests, whereas the subsequent drafting in of '*reasonably practicable*' provisions to the second Act dramatically altered the landscape. Business interests were protected; in this case ensuring the continued, albeit modified, 'right to pollute'. The move from protecting public (and planet) to business interests was the shift from protection of public concerns to protection of private interests. Corporate profit was therefore valued over people.

In 1874 the pendulum had swung in the opposite direction and the train changed tracks. The implications were enormous; not only in terms of pollution regulation in the UK. As the first large-scale legislative control of industrial pollution, its influence reverberated across time and in due course across the world, influencing pollution laws in many nations. Legislation to enforce the prohibition of damaging practices to protect the greater good had been successfully smothered by the industry lobby. It may have seemed like a small victory at the time, but in fact it was huge. It was to set the approach taken not only by governments but also those appealing for remedy. Radical and rapid change was replaced by an incremental approach to preventative laws. Piecemeal and compromised, they were the slow creep towards ever more damage and destruction. A hundred years later, when environmental legislation started to take shape in America, the Alkali Act of 1874 more than any other, had already shaped the consciousness of legislators, politicians and industry.

NATURE IN DECLINE

Let us backtrack to 1866. The first Alkali Act had been in place three years and had almost completely eliminated emissions from alkali works, much to the relief of nearby farmers and workers. The ineffectual Sanitary Act of 1866, on the other hand, promised only to fine factory owners whose chimneys emitted black smoke. It was an ominous omen for the future policing of pollution. Many prosecutions failed when faced with the 'shades of grey' defence: although their smoke was indeed dark, it was not actually black. Even when successful, the resulting fines were nominal, often no more than half a pound sterling. Unhampered by law, the use of coal continued to escalate unabated and the pea soup smog of London became thicker and darker.

Jevons' Paradox

The year was 1865: *The Coal Question: An Inquiry Concerning the Progress of the Nation, and the Probable Exhaustion of Our Coal-Mines* had just been published. William Stanley Jevons (1835–82), an English Economist, examined the UK's reliance on coal and questioned the sustainability of reliance on a finite, non-renewable energy resource: *'Are we wise,'* he asked rhetorically, *'in allowing the commerce of this country to rise beyond the point at which we can long maintain it?'*

Jevons' central thesis was that the UK's supremacy over global affairs was transitory, given the finite nature of its primary energy resource. In propounding his thesis, Jevons covered a range of issues central to survival, from: limits to growth, overpopulation, overshoot, post-global relocalization, energy return on energy input, taxation of the energy resource, renewable energy alternatives, to resource peaking (this last subject widely discussed today under the rubric of peak oil). He was to become famous for predicting that Britain would run out of coal within decades (it peaked in 1913). He claimed that the nation's industrial and imperial ascendancy came not so much from hard work and sound government as from its coal. Crucially, Jevons explained that improving energy efficiency typically reduced energy costs and thereby increased rather than decreased energy use, an effect now known as Jevons' paradox and exemplified in many areas of manufacturing, car use being the most obvious one. Increase of efficiency results in increase of use. Policy based on imposing energy efficiency targets to reduce use has precisely the opposite effect; use escalates.

Identifying one of the key weaknesses of reliance on fossil fuel as being non-renewable and pre-empting many of modern day arguments surrounding peak oil, Jevons used a simple but effective analogy to explain the consequences of collapse:

> A farm, however far pushed, will under proper cultivation continue to yield forever a constant crop. But in a mine there is no reproduction, and the produce once pushed to the utmost will soon begin to fail and sink towards zero. So far then as our wealth and progress depend upon the superior command of coal we must not only stop – we must go back. [6]

Like a lobster that is lured into a complacent slumber who eventually dies as the pot of water slowly heats, so the masses failed to comprehend the growing extent of the damage and destruction to their health and the health of the environment around them. It takes effort and determination to speak on behalf of the masses, so often so much easier to be lulled into accepting the prevailing norm even when the norm is set on a destructive trajectory. Few spoke out on behalf of the masses. Yet, recognition of the natural world in decline was tangible; trees were encrusted with soot and stunted by acid rain, the sunshine was clouded by grey particle filled skies that stretched as far as the eye could see and spread wide over the countryside. It was the beginning of a recognition of the trans-boundary effect of pollution; the impact of coal smoke did not stop at the city edges, affecting many more than just those within the area of use. Only the privileged could buy better air by moving upwind from centres of population and industry.[7]

In response to 19th century industrial Britain, a 'back-to-nature' movement was born. Intellectuals such as John Ruskin, William Morris and Edward Carpenter, who were all against consumerism, advocated that pollution and other activities were harmful to the natural world. Their ideas in turn inspired various proto-environmental groups in the UK, including the Commons Preservation Society, the Kyrle Society, the Royal Society for the Protection of Birds and the Garden City movement, as well as encouraging the Socialist League and the Clarion movement to advocate measures of nature conservation.

Anti-smoke proponents targeted the smokestacks of industry, a potent symbol of the damage and destruction their belching fumes were causing. Industrialists retaliated by pointing the finger the other way, blaming home use – unfettered as it was from the mandatory restrictions that were already governing factory use. Any further restrictions, they complained, would be unfair and impossible to achieve. The Smoke Abatement Society was set up in London in 1898 with its own smoke inspector, and very quickly the idea took hold with similarly named groups in other cities coming together to fight against the undesirable fumes. In the USA, inspired by the activities of their counterparts in the UK, Smoke Abatement Leagues sprang up to focus on the health, cleanliness, aesthetic, and moral implications of smoke. In New York, the Anti-Smoke League led a crusade against the smoke cloud developing over Manhattan. As a result of public pressure and mounting evidence, the city launched dozens of cases against smoke offenders in the early months of 1906, with members of the league notifying Health Department officials of offending stacks so that the city might initiate legal proceedings.[8] Upon reaching trial, witnesses to the offending stacks were located by the league to testify on the detriment the smoke caused them.

The pro-active approach of the Anti-Smoke League made a significant difference. The frequent daily arrests had their effect; within a remarkably short period there were very few chimneys left belching black smoke. This was single-issue activism at its most successful; by keeping the smoke problem before the public, municipal action was guaranteed.

Britain's pressure groups also kept the smoke problem alive in the public arena, pressurizing for new fuel solutions. Frequent favourable coverage in the *Lancet* helped the group evaluate the claims of smoke-preventing fireplaces. Reformers set up education programmes to advance the benefits of clean air and promote gas cooking and heating appliances. Even with local support, some hurdles were insurmountable without stricter enforcement; some factories were owned by the very officials whose duty it was to enforce the smoke laws. Stringent application of the law by local authorities was rarely applied through fear of driving away industries and jobs. As was well understood at the time, until environmental standards were enforced uniformly throughout the country, polluting factory owners would continue flouting the laws and continue threatening to relocate whenever local authorities tried to make them reduce their smoke output. Fines in any event were modest, and some factories were content to pay, if and when successfully prosecuted, rather than invest in more efficient means. Uniform standards and unbiased enforcement simply did not exist.

Smoke activists mobilized in cities across the country, forming the Smoke Abatement League of Great Britain in 1909. The following year a deputation met with the president of the Local Government Board. They argued that the only way to solve the smoke problem was through more active government involvement. William B. Smith, the acting president of the Glasgow and West of Scotland branch of the league, repeated a common refrain among reformers, namely that each locality *'could only take action with regard to works within its own area, whilst it often happened that there were works just beyond the city or borough which poured out smoke continually, yet no action could be taken.'*[9]

An international conference was held in 1911 and American reformers joined their British counterparts. Information, reports and scientific data were shared and disseminated. Conversations about strategy, tactics and insights stoked the fire of reform and change. Industry argued for efficiency and technological fixes. Jevons' paradox was quietly ignored and engineers were put to the task of researching and inventing more efficient installations. The government faced two choices; ban the use of coal or impose further restrictions. Yet again the choice was between protecting public or private interests. It was the concern of a few for the public wellbeing that had galvanized so much activity. Something had to be done, and the government knew it. So the government stalled for time and proposed an enquiry.

An enquiry was indeed set up in 1913, but external events played their hand. The intervention of World War I brought the death of thousands, which curbed the appetite for the fight against coal. Concerns about coal were put on the backburner as more immediate concerns took precedence.

It took death in peace-time to finally halt the visible pollution of the skies. The Great London Smog of 1952 still holds a strong memory for those who experienced it. The very high use of coal due to the cold weather and a lack of wind to disperse the toxic smoke led to a deadly cocktail of the deepest and darkest of smogs. It hung silently over London for just four days. Although not considered unduly alarming at the time, it proved for some to be deadly. Within the weeks that followed 4,000 deaths were recorded as a direct result of the air pollution. The final death toll reached 12,000. Finally it was understood that the burning of coal was lethal. It took the deaths of thousands finally to trigger a decision to stop the pollution. In 1956 the Clean Air Act was enforced and London's pea soup smogs were relegated to history. One of the outcomes was the designation of Smoke Control Areas where the use of certain fuel burning is either prohibited or only allowed in special appliances. Subsidies were given to householders to cover costs of adapting fireplaces in private dwellings, grants made available to churches and charities. Coal-filled boilers in factories and household coal burning fireplaces were replaced with gas and electricity, boats and trains switched to diesel engines. The coal mines closed: from over a million men working in the industry in 1913, the numbers had reduced to 10,000 by the end of the 20th century. [10]

On closer examination, what seemed like a victory for the Smoke Abatement Leagues was in truth very limited. Significantly, the use of coal itself was not fully prohibited, just where it was used and how. The Clean Air Act legislated zones where smokeless (i.e. non-visible) fuel had to be burnt and relocated power stations (that were exempt) to rural areas. The Clean Air Act of 1968 introduced the use of tall chimneys to disperse air pollution for industries that were allowed to continue burning coal, liquid or gaseous fuels. The problem had not been fully understood, and thus the problem had not been fully eradicated. The problem as it was understood to be, namely the smog, was deemed fixed. But smog was just a tangible outcome of the use of coal itself. To be truly rid of pollution, the source had to be correctly identified and prohibited – at source.

The prohibition of dark smoke had a dramatic impact that was immediate and tangible. For instance, vulnerable individuals such as patients in hospitals witnessed a direct and measurable improvement in terms of health and safety. Smoke from furnaces was crucially required only to *'be so far as practicable'* smokeless. The compromise terminology of the noxious gases Alkali Act (as further amended in 1906) was deployed. The *'reasonably practicable'* provisions and defence, whilst spawning a Clean Air Council and a whole new industry to handle Health & Safety and measurement requirements, once again sidestepped prohibition upstream. The Clean Air Act of 1968, after consulting with industry, firmly embedded the premise that smoke pollution should be controlled (raising the height of chimneys) not eliminated, by deeming it generally impracticable to remove sulphur dioxide. Industry's 'right to pollute' remained securely in place. The reality was that the pollution had been rendered invisible and displaced to elsewhere; out of sight and out of mind. With the advent of 'clean coal', the door remained open for the acceptable and continued use of coal, not just in the UK but throughout the world. An opportunity to usher in the changes was missed.

Pollution can be both visible and invisible. Coal – soot and grime – was viewed as a visible pollution that had to stop. Coal combustion was considered harmful primarily because of the unoxidised particles it produced.[11] Make the combustion process more efficient, it was reasoned, and the pollution would disappear. In other words, clean up the sooty particles and coal could still be used. If no visible pollution could be seen, the problem would, it was reasoned, cease to exist. The focus very soon shifted to the fix, not the need for change.

OUT OF SIGHT OUT OF MIND.

Coal is not just a hydrocarbon, breaking down into hydrogen and carbon dioxide when burned; it is a complex compound. It also contains sulphur, arsenic and mercury. Poisonous substances that do not disappear when burned. Instead they combine to create toxic acid that corrodes all it touches, plants, trees, stone, iron and our lungs. These substances have to go somewhere; either they remain in the ash or they go up the chimney. Taller chimney-stacks certainly did not solve the problem, as we were soon to discover. All this did was displace the toxins into the atmosphere at a higher level, where it travelled across national boundaries, falling eventually on Northern Europe. What disappeared up the chimney literally came back to haunt us, causing disputes over acid rain between the UK and Germany in the 1980s. It is little realized that we continue to burn mountains of coal, albeit no longer coal dug from our own mines. 60 million tons of coal is imported into the UK on an annual basis today and burned in centralised power stations, accounting for a third of our electricity supply.

The Clean Air Act, hailed as a milestone in the history of pollution prevention, marked a point in time where we could have shifted tracks. Instead, all we did was remove a few carriages, clean the windows, put in some filters and continue down the same route. This time we are expelling invisible pollution, much much more of it.

Chapter 2

MASSACRE OF
THE INNOCENTS

Alcohol Tax: The Indirect Oil Subsidy

Alcohol was not only imbibed for pleasure in the 19th century but more crucially it was the mainstay of providing light in America, replacing the once predominant but smelly whale oil. [1] Whale oil had fallen into decline by the 1850s, in part due to commercial whaling activities leading to the near extinction of the glorious leviathans of the sea, and in part due to the increasing loss of whaling ships being frozen and crushed in the Arctic ice. A far cheaper lighting fluid – and certainly more pleasing to the olfactory senses – was camphene. Kerosene derived from coal oil was sooty and smelly, and so camphene rapidly became the favoured fuel to light the streets and homes at night. Camphene was a mix of ethanol, turpentine and camphor. Ethanol is a clear, colourless chemical compound made from the sugars found in crops such as corn, sugar beets and sugar cane. It was a biofuel, and the capacity to grow ever more made it an attractive alternative. In the 1850s nearly 90 million gallons of ethanol were produced every year by distilleries in the US, with over 40% being used for lighting. By the 1860s camphene was the favoured lighting fluid, not only because of its sweet aroma, but also for its low cost. Whale oil had by this time risen steadily to over $2.00 a gallon, whereas camphene burning fluid was a quarter of the price. [2] The economic and aromatic advantage of ethanol over whale oil was not, however, to last for long.

War is a costly business and the American Civil War (1861–1865) was proving to be no exception. In order to cover the debt of war a swathe of tax reforms were imposed. In 1862 Abraham Lincoln passed the Revenue Act, introducing income tax and imposing taxes on various commodities including alcohol. The alcohol tax began at 20 cents and had risen to $2.00 per gallon just over two years later. Oil had been discovered in Pennsylvania in 1859 but it was not considered a serious contender to coal at that time. It hadn't been included in the list of war taxation by Lincoln, and indeed in 1862 its use was nominal.

Although most of the taxes declined in the aftermath of the war, post 1868 only two taxes were retained as the main source of Government revenue. For the next 45 years almost 90 percent of all revenue was collected from alcohol and tobacco as the remaining excises.[3] It was a tax that inadvertently provided an advantage and an opening for a new fuel to enter straight into the market at a favourable price point; that fuel was oil.

100 Years of Oil

After a shoot-out that left him with one arm and on trial for murder, Pat decided to move elsewhere for his fortunes. Pat was a prospector. He went off to study the latest exploration techniques, learning as he went about geology and mechanical engineering. Everyone was talking about oil, but the reality was very little had been found.

Pat knew the stories from his youth of dancing ghosts on nearby Spindletop Hill. He had a hunch that they were in fact fumes and heavy mist rising from sulphur gas escaping from below. Something of great worth, he surmised, could be lying underneath that very hill. A huge oil reservoir could just be waiting to be discovered.

Pat had daring plans. He wanted to build an industrial town, to be called Gladys City, filled with industries which would be fuelled by oil and gas from the hill. So, he formed a company and, with a small group, went back to his hometown and started drilling. Each hole was dry. So too were his backers who soon lost interest. Pat complained that the drilling rigs used by the contractors were too light. He wanted to drill deeper but couldn't without finance. Pat was laughed at and called a dreamer but finally he found a new backer prepared to take a chance. Captain Anthony F. Lucas set up business with Pat bringing in a few others; arguments soon followed and Pat left. Operations proceeded without him.

On January 10th 1901, two men working on the hole heard a rumble. It was 10.30 in the morning and mud spluttered out throwing them both back. Pat had insisted that the well depth had to be 310m, which was precisely the depth they had reached. Suddenly the earth roared and out shot oil at such speed that it reached a height of nearly fifty metres covering everything in its wake; people from miles around came rushing to dance beneath it. The flow was enormous and was estimated at 100,000 barrels per day.[4] The oil spurted at such incredible speed, energy and force that the skies were blackened for weeks. It was a flow rate greater than all of the oil wells in the United States and beat Drake well in Pennsylvania – the fist oil well – many thousand times over. In the first year of operation, the oilfield churned out over 3 million barrels (480,000 m^3), and over 17 million barrels (2,700,000 m^3) the following year. The Spindletop oil field was to continue operating for another hundred years.

Before the age of forty, Pat was to change the course of history. The oil industry was born. Spindletop became the first major oil field and one of the largest in American history. In just a matter of months Beaumont's population grew from 9,000 to 30,000 by March 1901.[5] Pat's wildest dreams were surpassed – the entire state of Texas prospered from the founding of a whole new industry. Pat's hunch brought him and thousands of other newly created companies to enormous prosperity: forerunners of many of today's major oil corporations – Texas Company (now Texaco), Sun Oil Company and Humble Oil Company (now Exxon Mobil) – can be traced back to the success of Pat's discovery.

Oil rapidly became a catalyst for many industries; petroleum refining and distribution, automotive, chemical and electrical, to name a few. The American discovery of oil marked the moment in history where technological leadership shifted from Britain to the United States. While the previous age of mechanization had spread the idea of splitting every process into a sequence, this was ended by the introduction of the instant speed of electricity that brought simultaneity. Electricity generated from oil opened the door to industrial mass production. Crucially, oil was not subject to heavy taxation and so provided an economical fuel upon which new mass-produced vehicles could run on. Where supply of oil escalated, the automobile industry flourished. Despite the remarkable size of the oilfield at Spindletop, it could not keep up with America's growing use of gasoline. Registered vehicles in the United States increased dramatically overnight and the first large-scale production line was up and running by 1902. The following year Henry Ford was producing cars in the thousands. Within a decade automobiles had leapt from 8,000 to 460,000 and by 1915 over 2,330,000 were in use.

Pat's discovery changed everything. No oilfield in America has ever surpassed the size and duration of Spindletop, but more than that, it gave birth to some of the largest industrial players in the world. The man who refused to let go of his dream set in motion events that heralded the beginning of the race to exploit our resources. His full name, somewhat disconcertingly, was Patillo Higgins.

THE CORPORATION AS A FICTIONAL PERSON

Within just a matter of a few years, industrialization in Texas had created the first giant industrial corporations with burgeoning global interests. Companies such as U.S. Steel and General Electric were joined by automobile manufacturers, railroad companies and the oil industry on the world's stock markets. It was during this time that the concept of a company being a 'legal person' took root. The starting point was the principle that a 'legal person' was simply an organization of human individuals. It is in fact correct to say that a corporation, consisting as it does of a group of individuals who have made an agreement as set out in their Articles or Charters of Association, has a legal written *agreement*.

A corporation is a community, of changing individuals who are working together governed by an agreed purpose. Literally, the people within a business are in unity over a common purpose. The roots of community are to be found in late 14th century Latin 'communis', a noun meaning 'common property, state, commonwealth.' In the late 15th century the word 'common' evolved in use to mean 'land held in common.' Unity derives from 13th century Latin *'unitatem'* (nom. *unitas*) denoting 'oneness, sameness, agreement', from *'unus'* meaning 'one.' By accepting the corporation as a community, the true relationship is expressed. It is the written documents of agreement that bind their relationship under the banner of a 'corporation.'

Prior to 1886, most companies were what are now called nonprofit organizations; the for-profit corporations were predominantly banks. At that time, and still in practice today, US charters were issued by states, each under its own laws. Early charters specified the corporation's purpose and expired at the end of a set term. If corporations overstepped their boundaries their charters could be – and not infrequently were – revoked. Corporations could not own shares in other corporations, so mergers and acquisitions and subsidiaries were unknown. Levying campaign donations to curry political favour was unlawful and statutes prohibiting corporations from contributing to political causes could be found in most states and the federal code.[6]

The turning point was a seemingly inconsequential case heard in 1886. A court reporter recorded comments made by the Chief Justice before the case of *Santa Clara County v. Southern Pacific Railroad* proceeded to trial. The words spoken by the Chief Justice Morrison R. Waite were not spoken as judgment, were not challenged nor subjected to legal examination. Yet, they were relied upon and accepted as law to justify the imposition of rights and responsibilities upon a 'fictional person'.

The Chief Justice informed the attorneys of the case before the trial commenced that

> *the court does not wish to hear argument on the question whether the provision in the Fourteenth Amendment to the Constitution, which forbids a State to deny any person within its jurisdiction the equal protection of the laws, applies to these corporations. We are all of the opinion that it does.*

It was the first time that the word 'person' in the Fourteenth Amendment was presumed to include corporations. Lawyers relied on this pre-trial comment to argue thereafter that corporations, or for that matter any legal entity, have the same constitutional rights and protection as a human being, as set out in the amendments and the body of the constitution itself. The outcome was dramatic. Justice William O. Douglas wrote in 1949, *'the Santa Clara case becomes one of the most momentous of all our decisions. [...] Corporations were now armed with constitutional prerogatives.'*[7] No longer could the state determine the behaviour and responsibilities of the corporate body.

The legal concept of the corporation as a 'fictional person' had in fact been around since the 17th century. The artificial person was used because there were certain resemblances, in law, between a natural person and corporations. Both could be parties in a lawsuit; both could be taxed; both could be constrained by law. But the idea that the fictional person had the same rights as a natural person was new. Reliance on the Fourteenth Amendment, one of the Reconstruction Amendments, to establish rights for 'fictional persons', was arguably invalid and morally wrong, as is demonstrated by close analysis of the Amendments. The Reconstruction Amendments are the Thirteenth, Fourteenth, and Fifteenth amendments to the United States Constitution, adopted between 1865 and 1870, the five years immediately following the Civil War. This group of Amendments is sometimes referred to as the 'Civil War Amendments' or the 'Three Reconstruction Era Amendments'.

The Amendments were intended to restructure the United States from a country that was (in Abraham Lincoln's words) *'half slave and half free'* to one in which the constitutionally guaranteed *'blessings of liberty'* would be extended to the entire male populace, including the former slaves and their descendants. No intent can be imputed from Lincoln's words to extend the freedoms even further to include corporate entities.

The Thirteenth Amendment (both proposed and ratified in 1865) abolished slavery. The Fourteenth Amendment (proposed in 1866 and ratified in 1868) included the Privileges or Immunities Clause, Due Process and Equal Protection Clauses. The Fifteenth Amendment, (proposed in 1869 and ratified in 1870) grants voting rights regardless of *'race, color, or previous condition of servitude'*. President Abraham Lincoln, an outspoken opponent of the expansion of slavery, did not live to see the Amendments being implemented, nor their subsequent application.

The Civil Rights Act of 1866 preceded the Fourteenth Amendment. Examination of the terminology used in the original Act sets out the correct basis of application. On March 13th 1866 Congress ordered: That the bill (S. 61)

> *to protect all persons in the United States in their civil rights, and furnish the means of their vindication, – be printed, together with the amendments of the House of Representatives thereto, and that the same be referred to the Committee on the Judiciary.*

AN ACT

To protect all persons in the United States in their civil rights, and furnish the means of their vindication.

Be it enacted by the Senate and House of Representatives of the United States of America in Congress assembled,

That all persons born in the United States and not subject to any foreign power, excluding Indians not taxed, are hereby declared to be citizens of the United States; and such citizens, **of every race and color, without regard to any previous condition of slavery or involuntary servitude**, except as a punishment for crime whereof the party shall have been duly convicted, **shall have the same right**, in every State and Territory in the United States, to make and enforce contracts, to sue, be parties, and give evidence, to inherit, purchase, lease, sell, hold, and convey real and personal property, and to full and equal benefit of all laws and proceedings for the security of person and property, **as is enjoyed by white citizens**, and shall be subject to like punishment, pains, and penalties, and to none other, any law, statute, ordinance, regulation, or custom, to the contrary notwithstanding (emphasis added).

The language of the Act, and the subsequent Fourteenth Amendment, do not in themselves support the theory that it was passed for the additional benefit of corporations. As the bold text above demonstrates, the phrasing used in the Civil Rights Act of 1866 firmly locates the extension of human rights to non-white citizens and former slaves – in law, what is referred to as natural persons, not fictional persons.

Corporate personhood is known as a 'legal fiction,' the term used by lawyers to describe a fact assumed or created by courts. A legal fiction is one that is known to be untrue but serves a specific purpose, a means to an end. This particular fiction, the fiction of the corporate personhood, does not state that a corporation – an artificial body created on paper – has the same rights and powers that are given to living breathing humans by modern democratic governments. The widening of the definition of a 'fictional person' in the *Santa Clara* case was, for all intents and purposes a radical legal departure. It was to change everything.

There are points in history where everything turns on a pinhead. Maybe an acupuncture needle is more apt in this instance. At the base of an acupuncture needle is the tip which, when inserted into the relevant acupressure point on a meridian line, facilitates the release of blocked energy. An incisive insertion at just the right moment and place can provide a release to that which is blocked, and a surge of energy then floods through. In life, the metaphoric acupressure point is not always identified with such consummate skill, and indeed is sometimes unblocked unwittingly – but the outcome can be equally dramatic.

Santa Clara County v. Southern Pacific Railroad has proven to be one such acupressure point. The reported words of the Chief Justice were the seemingly innocuous needle that released the block of governmental control over corporate activities. Enormous amounts of human energy began to flow in a very different direction; into untrammeled innovation, investment and production that resulted in a vast surge of growth. Corporations, no longer restrained by governmental restrictions, suddenly had the potential to access unlimited investment which could lead to unlimited growth and profit. Suddenly corporations had rights, just like humans. It was the new found freedom to pursue profit without responsibility for the consequences.

Within fifty years corporate personhood had been accepted worldwide. The IMF (fiscal and monetary issues), the World Bank (financial and structural issues), and the ITO (international economic cooperation) were all founded on the implicit understanding and acceptance of the enormous and unrestricted power and potential for profit corporations now held.

Kenny's Story

During the First Gulf War (1990–1991), US and British troops fired 320 tonnes of depleted uranium (DU) weapons at Iraqi tanks.[8] Kenny Duncan was 35 and a driver with the 7th Tank Transporter Regiment. He helped move tanks destroyed by shells. He had no notion at that time that the destruction caused by the use of DU coated ammunition would also impact his health and that of his (as yet unborn) children.

The conflict itself only lasted less than a year, and Kenny was there for just a matter of months. On his return his wellbeing deteriorated rapidly and he was dismissed in 1993 by the MoD on ill-health grounds. The following year his wife gave birth to his first son, soon to be followed by two further children. All three children displayed symptoms similar to those suffered by some Iraqi children, including deformed toes, and low immune systems making them susceptible to asthma, hay fever and eczema.

Kenny's pay-off for his nine years service was reduced to £40 a week, half the full pension. Kenny decided to appeal. Thousands of veterans had complained of a multitude of symptoms, only to be informed that there was nothing wrong with them. The difference for Kenny was that he was one of sixteen British veterans who had their blood tested for a research study. The tests of Dr Albrecht Schott, a German biochemist, revealed chromosome aberrations caused by ionising radiation. The men had fourteen times the usual level of chromosome abnormalities in their genes, raising fears that they would pass cancers and genetic illnesses to their offspring. These tests were decisive. Standard use of urine sample tests to determine impact of exposure to DU fail to examine gene abnormality.

The veterans tested had served time in the Gulf, Bosnia, and Kosovo. All three wars had deployed depleted uranium weapons. The Pension Appeal Tribunal Service hearing in Edinburgh accepted the medical evidence and Kenny was the first veteran to win a pension appeal after being diagnosed with depleted uranium poisoning during the 1991 Gulf War. [9]

In the aftermath of the war, Kenny's increasing breathlessness and aching joints were just the tip of the iceberg. DU contamination from tank battles raised numerous medical queries both in the UK and elsewhere; doctors in southern Iraq reported a marked increase in cancers and birth defects. A recent US government report[10] cites Gulf War Syndrome as adversely affecting one in four of all US veterans who served in the Gulf War, accounting for nearly two hundred thousand US army personnel. Hundreds of thousands of civilians who were near conflicts during, or downwind from the chemical weapons depot demolition after, the 1991 Gulf War are reported to have suffered from wide range of acute and chronic symptoms including (amongst other things) fatigue, loss of muscle control, headaches, dizziness and loss of balance, memory problems, muscle and joint pain, indigestion, skin problems, immune system problems and birth defects.

The 350 tonnes of depleted uranium ammunition left scattered across Kuwait and Iraq had its own consequences. After the Gulf War, Iraq was not allowed the equipment to clean up the battlefields. Air, water and soil were contaminated by DU dust, spreading for miles around. It was inhaled, ingested and digested. The inhabitants of the territory suffered, and continue to suffer, prolonged internal exposure leading to respiratory diseases, breakdown of the immune system, leukaemia, lung cancer and bone cancers. More than one million rounds of weapons coated in depleted uranium were fired on Iraqi tanks.[11] Kenny had unwittingly walked into a toxic soup.

The British Ministry of Defence (MoD) claim that Gulf War Syndrome does not exist. The National Gulf Veterans and Families Association (NGVFA) have asked the MoD to accept Gulf War Syndrome, and set up a separate pension. The MoD have to date refused to do so.[12] The 2008 US Gulf War Illness Report, commissioned by the US Government, takes a contrary view. The conclusions list as top priority a whole host of measures be put in place including: measures to distinguish veterans with Gulf War illness from healthy veterans; provisions for updated and ongoing information on overall and cause-specific mortality rates in Gulf War veterans and their families; additional analyses of Gulf War illness, cancer, respiratory conditions, and other health outcomes and the requirement for comprehensive information on family members of Gulf War veterans to include information on diagnosed conditions, multi-symptom illness, behavioural problems, and birth defects.[13] US government reports have routinely identified industrial pollution as a source of airborne particulates and other contaminants during the Gulf War.[14] The Illness Report is unusual in that it provided an opportunity to examine the effects of exposure to various chemicals, including depleted uranium, and the subsequent long-term hazards they pose on human health.[15]

In the 2000 Environmental Exposure Report, the US Department of Defence (DOD) Office of the Special Assistant for Gulf War Illnesses classified DU exposure into three low-concern categories[16] but, in line with all DU health risk assessments analysed by the Illness Report, it failed to provide insights directly related to questions concerning persistent symptomatic illness.[17] Long-term signs and symptoms of DU exposure, occurring from high intensity and/or long-term exposure, will wholly manifest only after an appreciable time lapse, several decades later. To date no laws have been put in place to regulate the use of depleted uranium and only one country has taken the decisive step to ban the use of uranium in their conventional weapons: Belgium.[18]

Depleted uranium is highly dangerous. If not stored correctly, dispersal of radioactive dust and releases of radioactive gas enter the air and water becomes contaminated. When uranium ore is processed, 85% of the radioactivity is left behind in the tailings (the waste by-products), and must be managed safely for hundreds of thousands of years.[19] Kazakhstan produces the largest share of uranium from mines (27% of world supply from mines), followed by Canada (20%) and Australia (16%).[20] Uranium is an essential heavy metal required for nuclear reactors. In order for nuclear fission to take place, uranium has to be 'enriched' in the U-235 isotope. During the enrichment process the fraction of U-235 is increased from its natural level (0.72% by mass) to between 2% and 94% by mass. The by-product after the enriched uranium is removed, containing a uranium mixture, is termed depleted uranium. Wherever uranium enrichment or DU occur, high levels of cloroflurocarbons are released. One market for this nuclear waste is to use it in the making of armaments to considerably enhance penetration: a warhead coated in DU will slice through a tank like a knife through butter. In Kosovo, US warplanes fired 31,000 rounds of DU ammunition during the Nato air campaign in 1999; in Iraq, US Air Force A-10 Tank-buster planes fired some 940,000 DU rounds during operation Desert Storm.[21] Many British veterans regard the failure to provide soldiers with information about the dangers posed by the use of DU was a breach of the duty of care owed to them by the Ministry of Defence, and Iraqis, who live or work near a modern battlefield where DU weapons remain, are likely to suffer even worse fates. In the UK, a report from Atomic Energy Technology, the commercial arm of the Atomic Energy Authority, prepared in 1991 finally became public in 1999 following parliamentary questions. It estimated that 500,000 extra cancers would occur in Southern Iraq over the next 10 years.[22]

It is not just in wartime that DU becomes airborne. Uranium

refineries pose an immediate danger for the residents of downwind communities. [23] Increased exposure to uranium can happen to those who live near or work in factories that process phosphate fertilizers, facilities that made or tested nuclear weapons, coal-fired power plants, facilities that mine or process uranium ore or enrich uranium for reactor fuel. Houses or structures built over uranium deposits (either natural or man-made slag deposits) or nuclear waste sites have an increased incidence of exposure to radon gas.

Just how much DU exists and how it is being stored is difficult to determine. The nuclear industry operates without transparency and accountability, therefore true figures are hard to come by. Likewise the armaments industry is not forthcoming on these figures.

On the 25th March 2010 the European Parliament's Foreign Affairs Committee recommended that the Council of Ministers (the principal decision making institution on Security and Foreign Policy matters of the European Union) support action on depleted uranium weapons at the 2010 UN General Assembly, stating: '(v) ...the need for effective arms control, including small arms and ammunitions containing depleted uranium, and to exercise its influence in support of wider, more practical and effective disarmament efforts and measures.' [24]

Without laws, little will be done to prevent the continued production of DU from nuclear waste and the escalation of a secondary industry developing and selling DU armaments. Without prohibition of nuclear armaments, the development, manufacturing and stockpiling of DU armoury will continue to escalate. But this is not enough: without banning the use of uranium, the source of the problem – nuclear waste – will never be halted. Nuclear waste dumping is one of the world's long-term unarticulated disasters of the future. To effectively stop an illegal market developing, legislation will be required to stop the creation at source. Until the use of uranium in nuclear reactors is prohibited throughout the world the problem will persist; DU will continue to be created and a willing secondary industry will always be ready to exploit that which is not dumped as waste for financial gain.

Since 1993, the US Department of Energy have handed over the responsibility for all domestic production of enriched uranium fuel to the US Enrichment Corporation (USEC). In that time USEC have racked up an environmental cleaning bill of over ten billion dollars, which has been paid by the taxpayer. USEC is now up for sale in the region of two billion dollars. By privatizing the USEC, the pursuit of profits to be made from uranium enrichment will be by necessity predominantly from the sale of DU. Since 1959, the US government has also limited the liability of nuclear utilities for damage caused by accidents. Until 1988, the utilities were only responsible for the first $560 million per accident. Now the limit stands at $7 billion, that is the same figure as the subsidy received by the nuclear industry annually. [25]

Weaponry used in the past 45 years of wars has caused a legacy that will outlast many generations. Whether or not destruction of human life is adjudicated to be lawful in time of conflict is set out in the laws of war. Consideration of destruction of the environment, however, arrived too late to stop the Vietnam War. 1961-1975 brought the onslaught of mass chemical warfare: Agents Orange, White and Blue were deployed extensively, causing widespread mutations, birth defects, cancer and death in humans, together with long-term environmental contamination, loss of forest habitat and wildlife. The chemical warfare proved to be so destructive to humans and their habitat as to be eventually beyond the limits of applicability. [26] This was for a number or reasons; a growing global environmental consciousness had taken root in the 1960–1970s, as had an increasing anti-war sentiment, both in the US and abroad. Additionally, these environmentally destructive tactics were played out in the first widely televised armed conflict.

Convention on the Prohibition of Military or any other Hostile Use of Environmental Modification Techniques (ENMOD) 1977, Articles 35(3) and 55(1)

The destruction of environmental life in Vietnam led to the creation of the ENMOD Convention and Articles 35(3) and 55 of

Additional Protocol I in 1977.

The ENMOD Convention Article I(1) prohibits states from engaging in '*military or any other hostile use of environmental modification techniques having widespread, long-lasting or severe effects as the means of destruction, damage or injury to any other State Party*'. Articles 35(3) and 55 in Additional Protocol I set out the prohibition and the duty of care provisions. This was the first law to set out explicit prohibition of mass destruction of the natural environment by the specific use of environmental modification (ENMOD) techniques during times of armed conflict. But, to date Articles 35(3) and 55 have never formally been invoked and no state party has ever claimed its breach.

Major military powers refused to be parties to the Additional Protocol I, including Iraq and the US. One of the reasons argued for refusing to sign was the absolute prohibition threshold; it made no allowances for the military to decide as and when they deemed it necessary to use such techniques. Various nations argued against the imposition of any obligations upon the army to protect the environment in armed conflict. It was a battle between protection of environmental interests and military freedom to destroy. The military's right to apply ENMOD techniques was not merely being restricted, but was to be banned. For many military powers, despite the knowledge of the enormous long-term destruction, this was a step too far.

The reluctance of major military powers to sacrifice what they perceive as an essential freedom of action is one of the reasons for their failure to provide any protection at all. War brings with it responsibilities, to be shouldered by all parties concerned. Failure by major powers to ratify ENMOD's Additional Protocol I mirrors their consummate failure to face up to their responsibility for damage and destruction of the environment.

Turning A Blind Eye To Waste

"There may be a link between earthquakes and disposal wells," said OKAGO (Oklahoma Oil and Gas Association) president Chad Warmington. "But we—industry, regulators, researchers, lawmakers or state residents—still don't know enough about how wastewater injection impacts Oklahoma's underground faults. We don't know enough about what's really going on in the subsurface to know how to mitigate some of this risk."

The Oklahoma Corporation Commission recently directed disposal well operators in "Areas Of Interest" to prove that disposal wells are not disposing below the Arbuckle formation and into the basement rock. What is "disposed of", although unseen, is a toxic mix. Polluted waterways and land can take months and sometimes years to come to the surface, whereas seismic activity occurs rapidly. Where there are disposal wells, it seems, earthquakes turn up too.

The largest registered earthquake in Oklahoma (2011) was a magnitude 5.7. This coincided with the beginning of shale gas exploration. It injured two people, destroyed 14 homes, toppled headstones, closed schools, and was felt in 17 states. It was preceded by a 4.7 foreshock the morning prior and followed by a 4.7 aftershock. The home of Joe and Mary Reneau (less than two miles from the epicentre of the quake) took six months to rebuild.

In 2013 Oklahoma recorded 109 earthquakes; in 2014 the number reached to 585 of a magnitude 3+. Earlier recordings dating back to 1972 give an average number of earthquakes as 2. Their dramatic increase in seismic events has attracted scientists from across the world.

(In the UK, where fracking has been given the green light, geology is believed be more heavily fractured and faulted than that of the US. Earthquakes from similar activity cannot be ruled out - indeed preliminary exploration has already caused seismic disturbances, as admitted by the industry.)

It's the disposal of the wastewater created during the process that is the causal link to the seismic activity. The wastewater is pumped down into the Earth (where there are faults). That way, it is hoped, it shan't return to the surface. But, just like dynamite, fracking wastewater can trigger something far worse. By lubricating the Earth's plates found at fault lines, plates can move more freely which in turn causes more frequent earthquakes. Seismologists confirm that the disposal of water into or in communication with basement rock presents a potential risk for triggering seismicity.

For the state of Oklahoma and for many homeowners there, this is not good news. The state is now having an average of 2.5 earthquakes of at least magnitude of 3+ *every day*; it used to average less than two earthquakes *a year*.

Even Oklahoma's local government can no longer turn a blind eye to the mounting evidence – their own website has posted up The Oklahoma Geological Survey [27]. In a summary of it's findings, it states that the majority of recent earthquakes in central and north-central Oklahoma are likely triggered by the injection of produced water in disposal wells. The site very helpfully includes an interactive map that displays quantity and distribution of earthquakes; in areas where drilling operations have been disposing of fracking wastewater you will see clusters of seismic activity. Scientific consensus, that the earthquakes now rocking the state are largely caused by the underground dumping of billions of barrels of wastewater from fracking, has now been accepted by Oklahoma's government – some might say too late.

How we deal with waste can be approached in a number of ways; go upstream and make provision from the outset so that little or none is produced (and that which is, can be responsibly dealt with at source), or downstream and be faced with an ever rising pile-up (which is frequently dumped). The cumulative impact downstream is all too often hidden – shipped off, buried or poured underground.

Any process that involves enormous amounts of fluids that contain – even at a very small level – substances, which when combined can result in adverse environmental and human health impacts, has consequences. Choose fracking for instance, and further down the line dumping of millions of gallons of wastewater – as experienced in Oklahoma – triggers even more harm. Alternatively, choosing to put in place systems that are non-harmful, that do not depend on extractive methods, consequentially the cumulative impact of daily harm does not occur.

Calculation of what constitutes significant harm can be misleading. For instance, by combining high and low readings, an average figure is all too often accepted as low enough to satisfy concerns. Kenny, our successful depleted uranium pension claimant, could have lost his case if it had been determined on this basis. When walking into a war zone he may or may not be exposed to DU. If measurements of mixed high and low readings of DU are treated as an average reading, some may suggest that his general exposure would be inconsequential. His tribunal, on the evidence presented, clearly thought otherwise. An additional problem was that he was never informed of the potential long-term risks. Like Kenny, many may be adversely affected, just by dint of stepping onto the site. It may be many years until we have full understanding of the consequences and long-term implications for all who visit or use the Olympic site over the next two years. Short-term cost saving by the ODA has come at the expense of long-term protection for the users of the land. Because limited duration of exposure has been determined as acceptable, all other concerns have been swept under the carpet. Low risk assessment, premised on averaging out the numbers, has been sufficient to deflect the burden of paying out the additional costs of long-term rehabilitation.

Examination of large-scale investment by fund-managers and banks into corporate activity reveals that an enormous amount of finance is invested in damaging and destructive practices. Governments sanction enormous subsidies, often to sustain business practices that are the most destructive of all. The citizen pays taxes that are then redistributed; you pay money into your pension fund that is then invested, you pay money into your savings that are used to bank-roll other ventures. Where and how this money goes is rarely known, questioned or understood. Once out of your hands, banks, governments and pension providers determine where your money will flow.

Royal Bank of Scotland: A Lesson In Progress

In the autumn of 2007 Royal Bank of Scotland (RBS) announced profits of £10.3 billion, the biggest ever for a Scottish company and the equivalent of £1 million per hour. It was hailed as Scotland's own economic miracle; it was the fifth largest bank in the world. It did it by borrowing billions to buy 26 other companies in the space of seven years. Such rampant uncontrolled purchasing could not, in financial terms, be sustained and as a consequence within a year it all went spectacularly wrong. In the wake of the 2008 banking crisis RBS losses were recorded as £28 billion; the biggest in British corporate history.

Mervyn King, governor to the Bank of England, revealed that RBS received £36.6bn from the government to prevent it from collapsing in October 2008. Following a second round of recapitalization at £20bn in November 2009, the UK taxpayer now owns 84% of the bank's shares. Public ownership of the bank means that the bank has to be accountable to the public for the way our money is being used by the bank. An investigation by *The Guardian* showed that in the first six months following the bank's initial recapitalization in October 2008, RBS had been involved in loans worth nearly £10 billion to oil, coal and gas companies – a quarter of the total amount of public funds put into RBS at that point. [28] Prior to the bail-out, RBS had approved a $2.5 billion loan to the energy corporation Conoco Phillips for tar sands financing; [29] the money used to pay for. that loan was the British taxpayers'. In an open letter to the then Chancellor Alistair Darling, 30 public figures, including MPs, faith leaders and members of the business community, called on the Treasury to govern the bank; *'the government has effectively written a blank cheque for the rescued banks to finance anything from destructive fossil fuel companies driving climate change to hostile take-overs that threaten UK jobs'.* [30]

RBS is a people's bank, quite literally. The people, those who have paid their taxes, have the right to have a say in what that money is used for. When Stephen Hester, CEO of RBS was challenged on this issue at RBS's AGM in Edinburgh April 2010 and specifically asked why it was that he invested millions of the taxpayers' money into environmentally destructive projects, his response was *'well, it's not a crime.'*

RBS has invested millions of the taxpayers' money into the most destructive industry of all; fossil fuel extraction. Unless environmentally high risk and destructive activities are considered a crime, banks, investors and governments will continue to support business projects and investments that are on a steep trajectory to destroy the planet.

On July 2nd 2010, in response to the UK financial crisis, The Financial Reporting Council (the FRC) published a voluntary UK Stewardship Code imposing a duty of care on institutional investors.[31] In a similar vein as the UK Corporate Governance Code, the Stewardship Code is to be applied on a 'comply or explain' basis. Reports by institutional investors 'should' include a description of how the Code has been applied and disclose specific information. In the event that this is not done, institutional investors are requested to provide an explanation. There is no legal requirement to do so and in the event of noncompliance, there is nothing the FRC can do to enforce the code. There is no definition of what stewardship is and compliance is voluntary. Governance of this kind in response to a crisis that triggered the collapse of the UK's largest bank is weak at best and completely fails to govern the bank's own investment policies.

The Code applies to asset managers (pension funds, insurance companies, investment trusts and other collective investment vehicles), foreign investors, proxy voting and other advisory services. From the end of September 2010, the FRC will maintain on its website a list of all investors which have published a statement on their compliance or otherwise. The Code contains seven main principles and associated guidance. The main principles of the Code are:

1. Institutional investors should publicly disclose their policy on how they will discharge their stewardship responsibilities;

2. Institutional investors should have a robust policy on managing conflicts of interest in relation to stewardship and this policy should be publicly disclosed;

3. Institutional investors should monitor their investee companies;

4. Institutional investors should establish clear guidelines on when and how they will escalate their activities as a method of protecting and enhancing shareholder value;

5. Institutional investors should be willing to act collectively with other investors where appropriate;

6. Institutional investors should have a clear policy on voting and disclosure of voting activity;

7. Institutional investors should report periodically on their stewardship and voting activities.

The UK government, in bailing out RBS, acted in their capacity as an institutional investor on behalf of the taxpayer. All of the above 'should' apply to any regime that administers large sums of money on behalf of the taxpayer; for it is the taxpayers who are the shareholders in the business being invested in. Financing decisions of RBS, despite the public being the primary shareholder, remain focused on destructive activities. From December 2008 to May 2009, another $2.473bn of the British taxpayers' money poured into the Athabasca tar sands,[32] the largest ever unconventional oil extraction project in the world, up to five times more carbon intensive, energy intensive and water intensive than any other form of oil extraction process.

Chapter 3

TELLING THE TRUTH ABOUT THE BIRDS AND THE BEES

The Dead Ducks of Athabasca

The rush to find more oil has taken exploration into unconventional territory. Now no longer able to extract gushing oil from a few hundred metres below the earth's surface, corporations exploit fossil fuel from ever more extreme territories. Unconventional tar sand extraction is the new frontier. Without any more easy access to high-grade oil one new route is to strip-mine it from underneath ancient peat bogs and boreal forests, where it exists as a heavy tarry substance called bitumen. One of the consequences of unconventional tar sand extraction is the scarring of a wilderness landscape: millions of tonnes of plant life and topsoil are gouged out to make way for vast open-pit mines. Millions of litres of water are diverted from rivers (up to five barrels of water are required to produce a single barrel of crude) and the process requires huge amounts of gas.

Producing crude oil from tar sands generates up to four times more carbon dioxide than conventional drilling. Tar sands are a heavy mixture of bitumen, water, sand and clay; a large amount is to be found beneath more than 54,000 square miles of prime forest and peat bog in Athabasca, northern Alberta in Canada. That is an area the size of England and Wales combined.

Each spring more than half of America's birds flock to the Canadian boreal forest to nest. Hundreds of thousands of birds are dependent on the wetland habitats in the tar sands, the Peace-Athabasca Delta and other parts of the Mackenzie River watershed. Changes to Alberta's rivers and underground reservoirs are already having profound impacts on the wildlife. Expansion of the tar sands into the 35 million acres of Boreal forest will destroy and fragment bird nesting habitat of up to 170 million birds who breed there annually, causing significant problems for birds as well as other wildlife. All four major flyways in North America – the aerial migration routes travelled by billions of birds each year – converge in one spot in Canada's boreal forest, the Peace-Athabasca Delta in northeastern Alberta. This is known as North America's premier nesting ground. Millions of birds, including tundra swans, snow geese and countless ducks, stop to rest and gather strength in these undisturbed wetlands, each autumn.[1] For many waterfowl, this area is their only nesting ground. Remarkably, despite the importance of this area, little of the land in Alberta has protected Bird Sanctuary status. Under the Canadian Migratory Birds Convention Act, a geographical area may be designated as a Migratory Bird Refuge. However, although this restricts activities targeting a specified set of birds in that area, it does not protect the land or water features.

Tar sands oil extraction in the boreal forest just south of the Peace-Athabasca Delta, including sites upstream on the Athabasca River, is having a direct impact on this territory. Water extracted for tar sands mining reduces flow into the delta, killing fish – a food source for birds – and disturbing habitat. Wastewater discharge contaminates the river, creating a toxic food web and leading to reproductive problems in wildlife.

In Spring 2008, a flock of 1,611 waterfowl landed in 1.8 mile wide Aurora North Site mine facility tailings pond, north of Fort McMurray. The birds were en route to nesting grounds when they landed. The tailings ponds stretch out for over 80 square miles, twinkling in the sun as far as the eye can see. For the birds it was a true siren's call: so alluring but deceptively deadly. Instead of the boreal forests and wetlands providing safe nesting habitat for songbirds and waterfowl as they have done for thousands of years, they flew into a toxic waste pond. Usually the oil companies have a preventative method in place to stop the birds from landing: air cannons. The cannons trigger intermittent loud shooting sounds. That morning they had not been turned on. 1,611 birds landed and either sank to the bottom of the pond after becoming coated in oily sludge or were shot. Only five ducks were saved.

Cannons were originally created for the aviation industry where, unlike farms that could spread nets over their produce, the intention is to prevent birds flying into planes on runways. Reed Joseph International pride themselves as being largest US distributor of complete bird and wild life abatement system. Their solution is named 'Scare Wars'; an arsenal in the ongoing war against the birds, integrating intermittent firing with distressed bird sounds. The canons are inexpensive, low maintenance and very loud; blaring out imitation gunshot and screeching birds in pain at very short intervals, between 12 and 16 blasts per minute.[2] It is rather akin to an alarm clock going off every few seconds, but instead of a shrill ring, the sound of war on birdlife ranges between 115 to 130 decibels: loud enough to be considered by some as seriously distressing.

In general, Canadian laws do not cover such eventualities as the extensive use of these machines to scare birds away from tailing ponds. A charge has been brought, however, under Canada's Migratory Birds Convention Act, 1994 (MBCA). It is a law that is used to prosecute hunters and companies that dump hazardous chemicals or oil into the water. Until now it has never been applied to a tailings pond operator. Syncrude, whose liquid tailings account for 132 billion gallons,[3] have been charged with one count under subsection 5.1(1) of the MBCA for allegedly depositing or permitting the deposit of a substance harmful to migratory birds in waters or an area frequented by birds. Environment Canada officers investigated the incident along with Alberta Environment and Alberta Sustainable Resource Development. This was the single largest reported incident of oiled birds in the oil sands region.

Robert White, the lawyer representing Syncrude, asserts that the company had done everything it could to keep birds away from the tailings pond. The company say a late winter storm prevented them from putting bird deterrents in place and the birds migrated earlier than usual.[4] If convicted, Syncrude faces fines of up to $800,000. A second plea of not guilty has been entered against the charge of failing to have appropriate deterrents in place at a tailings pond, contrary to section 155 of the Alberta Environmental Protection and Enhancement Act (EPEA) which states that a person who keeps, stores or transports a hazardous substance or pesticide shall do so in a manner that ensures that the hazardous substance does not directly or indirectly come into contact with or contaminate any animals, plants, food or drink. The maximum penalty for a violation under this section is $500,000.

The state government of Alberta state on their information website: *'Hundreds of thousands of water birds travel through the Fort McMurray oil sands areas. Effective bird deterrence is an important part of the Alberta government's approval requirements for tailings ponds.'*[5] The Alberta Government's primary concern is the effective functioning of the oil sands, not the death of wildlife.

Poisoning the Pests

Malathion is just one of thousands of pesticides on the market today. It was one of the earliest pesticides developed and was first introduced onto the commercial market in 1950. A wide-spectrum organophosphate (OP) insecticide, like all insecticides, its purpose is to kill insects. Chemically similar to compounds developed as a 'nerve gas' for proposed use as chemical weapons before and during the Second World War, malathion has a similar mode of action. Malathion is frequently used to kill insects that feast on fruits and vegetables by aerial spraying over large areas of farmland. Domestically, it is employed in concoctions and remedies for killing flies, household insects, animal parasites and head lice. Malaria eradication programmes have included airspraying of urban areas to kill mosquitoes. When it comes to killing insects this pesticide is enormously effective. Malathion is one of the most common organophosphate insecticides applied domestically and commercially in the United States, and is used extensively throughout the world. [6]

Testing for the side effects of malathion has historically been left to the chemical industry. In the United States, the Environmental Protection Agency (EPA) does not test pesticides. The EPA only sets up the guidelines for the testing. The chemical companies then conduct their own tests and present the results to the EPA for review. In essence, the industry self-regulates. Pesticide Action Network (PAN),[7] who rely on wider research, rate malathion as a Bad Actor Chemical. [8] Non-industry research contains a range of findings, including examination of human poisoning and death from malarial airspraying in Pakistan, intestinal disorders in children born after Californian crop spraying, child leukaemia and aplastic anaemia after malathion exposure, second generation health problems in animals and identification of symptoms of human immune system weaknesses. It is widely accepted that malathion becomes more toxic when stored for three months, is exposed to sunlight or when temperature increases, and is often used in sprays in its technical grade. [9]

Despite the extent of external testing of malathion, the EPA state they require more detailed tests for chemical effects upon the immune and nervous system before coming to further conclusions. These tests demonstrate that malathion can remain intact for exceptionally long periods (many years); can be widely distributed throughout the environment as a result of natural processes involving soil, water and, most notably, air; and are toxic to human and non-human life. Despite this, it is not considered a Persistent Organic Pollutant (POP) under the Stockholm Convention, and is not listed as one of the initial 'Dirty Dozen' banned chemicals. Radical in its approach, it took almost 10 years of negotiations to have the Stockholm Convention ratified by 160 countries in 2004. The then executive director of UNEP, Klaus Toepfer of Germany, took a hard-line – that POPs must be stopped. '*The Stockholm*

Convention will save lives and protect the natural environment – particularly the poorest communities and countries – by banning the production and use of some of the most toxic chemicals known to humankind,' Toepfer announced. In 2013, after further extensive negotiations and submissions of research, a few more chemicals were added to the list. Despite the enormous body of evidence malathion, like many others, is still largely unregulated.

Perhaps what is more disturbing about malathion (and other chemicals as well) is not what is actually known about the health effects from the chemical, but rather, what is not yet known. We do not yet know the full effects of the use of this chemical on other aspects such as increase of infertility, miscarriage and breakdown of autoimmune systems. These effects have been found to occur after exposure to other pesticides and chemicals when tested by various University Research Agencies. What is often overlooked is the fact that the amount of chemical exposure required to cause the above effects is many times lower than the amount required to cause cancer, birth defects, organ damage or other major effects.

Toxicity testing for individual chemicals is the accepted EPA standard. Governmental agencies list malathion as a chemical of 'moderate toxicity' but this labelling fails to reflect the true severity of exposure. One test proves to be instructive: malathion was given to chickens at levels of up to 10,000 ppm for 15 weeks. All of the birds died, but because the chemical was excreted through the urine (the standard testing to determine toxicity) and none showed microscopic evidence of nerve damage, determination of toxicity level remains as 'moderate'. Characteristic signs of chemical toxicity were not present, although death resulted. Insects however, do not provide urine samples and so their toxicity is assessed somewhat differently. Mortality rates determine whether or not the pesticide is effective, namely whether it has killed the predator. As a result malathion is judged to be 'moderately toxic' to birds, yet 'highly toxic' to honeybees. As with numerous other pesticides that are also deemed toxic to varying degrees, it is not a banned substance. Manufacturing and application of malathion continues unabated.

The whole process is a hermetically sealed circle: the EPA say more testing is required before they adjust their guidelines, the guidelines in the meantime remain as they are, so the testing for subtler and/or long-term effects are not forthcoming, and the results relied upon are the ones provided by the very industry who created the chemical in the first place. Responsibility for the effects of the chemical has been neatly sidestepped by industry and governments have failed to ensure responsibility is imposed by law.

Testing Toxicity to Bees

Billions of bees have been mysteriously dying throughout the world for a number of years now. The phenomenon has been named Colony Collapse Disorder (CCD). One species that has been particularly badly hit is the *apis mellifera*, more commonly referred to as the honeybee; these bees are the pollinators that bestow upon us a third of all the food that ultimately reaches our plates.

The World Organization for Animal Health says that no one single cause lies behind CCD, claiming its causes are distributed and varied and without recourse to a miracle cure. Scientists say they are no nearer to knowing what is causing this catastrophic collapse, but there is plenty of evidence that modern pesticides such as malathion have played their part. For the fourth year in a row, more than a third of honey bee colonies in the United States have failed to survive the winter. More than three million colonies in the U.S. have died and the fields continue to be sprayed with pesticides. The very substances that are destroying the harmonic cycle that produces our food, are sold on the premise that they will help provide food security.

LD50, one of the primary tests relied on to determine toxicity of chemicals on a given species, fails to investigate any relationship – or lack thereof – to Colony Collapse Disorder. LD50 measures only individual toxicity, not colony toxicity. It does not account for the secondary contamination pesticide exposure one bee can have on the colony. For example, a moderate to low toxicity pesticide (by LD50 measurement) used in granular form that is collected and concentrated along with pollen might have little toxicity to adult bees, but devastate the colony by its indirect effect on hive reproduction or mortality rate of larvae or young bees. No LD50 studies measure the effects on larvae, concentrating instead on singular adult bees. By ignoring the migratory nature of pesticides, they (according to LD50 measures) can be concluded to be 'safe' to use when applied to a location free from blooming flowers that attract bees.

Few people know about the Aarhus Convention, in part because very few authorities and governments have implemented adequate steps to allow the public to seek ecological justice. The Aarhus Convention was named after the city in Denmark where the final talks were held, it entered into force in 2001 and has to date 47 members (including the whole of the EU). Authorities must make 'appropriate provisions' for public participation in the preparation of plans and programmes 'related to the environment.' It establishes that sustainable development can be achieved only through the involvement of all stakeholders and that states are to provide an inexpensive, accessible forum in which justice can be accessed. The Convention links environmental rights and human rights and is legally binding on those states that have chosen to become Parties to it. It addresses relations between the public and public authorities in a democratic context and requires those authorities to consider the public's views when taking decisions, thereby linking government accountability and environmental protection.

Citizens now have the right to challenge government departments when they take decisions related to the implementation of the Aarhus Convention which maintains that we owe an obligation to future generations. The Convention's broad definition of 'environmental information' includes air, water, soil, and biological diversity, and the public has a role to play in decisions about activities – such as crop spraying – that could potentially be damaging.

Toxic Chemicals in Agriculture report, 1951

In the 1950s the British Ministry of Agriculture and Fisheries (now called DEFRA) first became involved in the control of pesticides. This followed the report of the Gowers Committee on Health, Welfare and Safety in Non-Industrial Employment, published in 1949[10] which revealed, amongst other things, a high level of fatal accidents involving the spraying of pesticides to kill weeds. An advisory committee on pesticides was set up, chaired by Professor Solly Zuckerman, a zoologist by training who knew little about agrochemicals and expressed surprise at his initial appointment. Writing a chapter of his memoirs entitled 'An Amateur in Whitehall' Zuckerman recalled of that committee: *'We had several discussions about the desirability of compulsory and statutory controls, but in the end decided that a voluntary arrangement with the industries concerned was the best that was then possible.'* [11]

The conclusions of the Toxic Chemicals in Agriculture report published in 1951 were conflicting and inconclusive. The committee's main recommendation was the setting up of another committee whose task would be to 'advise generally' on problems relating to consumer health. Voluntary arrangements with polluting industries were preferred over hard law. Recommendations included measures for 'adequate protection' of people using pesticides and that all containers should be worded: 'Deadly Poison.' Decisions as to what constituted too dangerous to be put in the public domain was to be left to the industry to determine.

COMPROMISING ON POISON

Rather than banning the poisons, the Agriculture (Poisonous Substances) Act 1952 was enacted, which empowered the Minister to make regulations to specify protective clothing to be worn, procedures to be followed and the minimum age of workers to be employed when using pesticides. In the end three successive official Working Parties were headed by Sir Solly Zuckerman, first on user-risks (H.M.S.O., 1951), then risks to the public health (H.M.S.O. 1953) and finally risks to wild life (H.M.S.O. 1955). No further laws were passed. Industry's rights to pollute and self-regulate were protected. Much of the chemical experimentation of the period was sponsored by the military; any further assessment of 'safe use' of agrochemicals was left up to the manufacturers for the next 30 years. The door to unfettered experimentation (and profits from poisoning) lay wide open.

In 1986, 35 years later, the first UK law to regulate pesticides (Control of Pesticide Regulations) was put in place. The UK policy decision was to *minimize* use, not to reduce or prohibit. However, unlike other counties, no limits were set. Pesticides have become increasingly sophisticated in any event, and the tonnage of active ingredient applied in the UK declined substantially not because of effective laws but because of increased use of products which were more biologically active at lower dosage rates than those they replaced. [12] Less chemicals are required to do an even better job than before. As Jevons noted 100 years before, increases in efficiency merely meant that prevalence of use increased exponentially. Nevertheless, it was the UK government's expectation that a policy of minimizing use would lead to an overall and sustainable decrease of use over time. Time has proven this not to be the case. It is estimated that approximately 31,000 tonnes of pesticide active substances are now used in the UK each year. 80 per cent of pesticides are used in agriculture and horticulture; 16 per cent are used in homes and gardens and 4 per cent in the amenity sector. [13]

European legislation introduced to address dangerous substances took an even longer time to be implemented in the UK. The Directive to Control Water Pollution by Discharges of Certain Dangerous Substances [14] was one of the first water related Directives on pollution to be adopted. The aim was to regulate aquatic pollution by the thousands of chemicals now being produced in Europe. Inland surface waters, territorial waters, inland coastal waters and ground water were being increasingly polluted and so a two tier listing of substances was drawn up of some of the most dangerous individual substances. Their emissions were to be limited (a form of capping) and minimum requirements were based on application of industry's 'best available techniques'. Each EU country had to set up its own reporting lists and requirements. The explicit purpose of the Directive is to eliminate pollution of a limited number of identified chemicals and reduce pollution from various other substances. Implementation across Europe was slow, to say the least. It was not until 1998 that the UK finally implemented its Surface Waters (Dangerous Substances) (Classification) Regulations. Included in its Pollution Inventory of substances of potential environmental concern of national significance is malathion.[15] It is termed a UK 'Red List' pollutant, denoting the presence of which in the environment is of particular concern.

Overall sales of pesticides in the UK alone are now in the region of £490 million every year. The UK government regulators receive approximately £7 million a year from the industry to grant pesticide approvals. Pesticide use across the world has increased 50-fold since 1950[16] and the value of the world pesticide industry grew a staggering 29% to US$52 billion in 2008. In America, the EPA stopped reporting pesticide use statistics in 2001.[17] Minimizing use of pesticides has become a global business in itself.

A small group of UK MPs believe that there is another way of dealing with pesticides that are organophosphates; namely by banning them. The All Party Organophosphate (OP) Group are calling for a *'moratorium on the use of all organophosphates until an accurate assessment is made of toxicity and the mechanism of damage, particularly in relation to long-term chronic effects as a result of cumulative exposure to OPs.'*[18] It's a small voice shouting 'stop' from the inside of our incredibly noisy train when all aboard seem to be hell-bent on pollution partying. They advocate a radical approach to stop the juggernaut-train: nothing less than laws to prohibit use will work.

The Wrong Kind of Evidence

Georgina Downs is on the outside pulling up the tracks. Founder of the UK Pesticides Campaign, for the past nine years Georgina has single-handedly taken on the UK government over the profligate use of pesticides and crop spraying. In November 2008 a High Court Judge ruled in Georgina's favour, upholding her claim against the UK government for its failure to protect rural residents against exposure to harmful pesticides. It was the first known legal case of its kind to directly challenge UK pesticide policy and agricultural practice.

Georgina argued that the 1986 Control of Pesticides Regulations provides for various safeguards to be put in place for beekeepers in advance of crop spraying, but no residents of the vicinity were afforded any protection whatsoever. Living on the edge of farmland near Chichester, West Sussex, Georgina was first exposed to pesticide spraying at the age of eleven, as a result she suffered extensive and long-term injury to her health. Georgina argued that the government had failed to address the concerns of people living in the countryside who are repeatedly exposed to mixtures of pesticides and other chemicals throughout every year, and in many cases like hers, for decades. People are not given prior notification about what was to be sprayed near their homes and gardens, in direct contrast to those who are spraying who by law must wear protective clothes and masks. Despite the explicit recognition of the hazards of use of the pesticides, nothing is done to protect the community who are exposed long-term to their use. Mr Justice Collins agreed with Georgina Downs' 'solid evidence' that people exposed to chemicals which are used to spray crops had suffered harm. The government immediately sought to appeal the judgment.

One year later the Court of Appeal overturned the decision, ruling that the detailed evidence Georgina had amassed over nine years could not be used against the government because she had 'no formal scientific or medical qualifications.' Georgina is now taking her case to the European Court of Human Rights.

Deteriorating Nutrients

In the 95 years since chemicals were introduced as a means of waging war against fellow human beings, we have seen the use of chemicals expand prolifically in the war against nature. Yields may have improved in the short-term, but nutritional values have dropped dramatically. According to the UK government's own data, between 1940 and 1991 the typical British potato "lost" 47% of its copper and 45% of its iron. Carrots lost 75% of their magnesium, and broccoli 75% of its calcium. The pattern was repeated for vitamins. A study in Canada showed that between 1951 and 1999, potatoes lost all of their vitamin A and 57% of their vitamin C, while today's consumers would have to eat as many as eight oranges to obtain the same amount of vitamin A their grandparents did from a single fruit. [19] Despite the enormous body of evidence dating back over 75 years, malathion and a whole host of newer chemicals are sprayed liberally over our food, more now than ever before. In California alone, 489,650 pounds of malathion were used in 2000. That is just one statistic of one area and the use of one chemical. [20] The bioaccumulative effect of the toxic cocktail of mixing chemicals has yet to be determined. It is a treble hit; the inherent nutrient value of our foodstuffs has been eroded, all species suffer long-term and now we are eradicating the lives of the very insects that ensure our foodstuffs grow.

Rachel Carson the biologist, writer and ecologist was the first major whistle blower. Chemical corporations vigorously attacked her carefully researched exposure of the environmental damage caused by widespread use of pesticides. The writer of *Silent Spring* (1962) stands as a truth-teller of exceptional courage and insight. Her words have all the greater relevance today than ever before:

'These sprays, dusts, and aerosols are now applied almost universally to farms, gardens, forests, and homes – nonselective chemicals that have the power to kill every insect, the 'good' and the 'bad,' to still the song of birds and the leaping of fish in the streams, to coat the leaves with a deadly film, and to linger on in soil – all this though the intended target may be only a few weeds or insects. Can anyone believe it is possible to lay down such a barrage of poisons on the surface of the earth without making it unfit for all life? They should not be called 'insecticides,' but 'biocides.'

Part 2

PROTECTING
OUR *OIKOS*

'We stand now where two roads diverge. But unlike the roads in Robert Frost's familiar poem, they are not equally fair. The road we have long been travelling is deceptively easy, a smooth superhighway on which we progress with great speed, but at its end lies disaster. The other fork of the road – the one less travelled by – offers our last, our only chance to reach a destination that assures the preservation of the earth.'

RACHEL CARSON, conservationist,
marine biologist, ecologist (1907–1964)

Chapter 4

PRINCIPLES TO PROTECT OUR *OIKOS*

Laws to Prevent, Pre-empt and Prohibit Ecocide

1. Amend all compromise treaties, laws, rules and regulations:
    (i) replace with prohibition of all damaging and destructive practices; and
    (ii) include provisions to enable restoration of damaged territories to be prioritized over existing practices that are premised on financial penalty alone;

Sacred Trust of Civilization

2. Community interests to be placed over private and corporate decisions;

Holding Business to Account

3. An elected community member of the territory concerned to be seated on a company's board of directors;

4. Accountability of business practices to be scrutinised by independent bodies and made publicly available; transparency of process to be open to public comment;

5. Where lobbying at a political level is undertaken by corporate entities, a record of all activities and monies spent to be made available for public scrutiny and query;

ENVIRONMENTAL SUSTAINABILITY

6. Hold all governing bodies to account. If damaging policies are being promoted, call for them to be halted with immediate effect;

7. Hold community meetings and ensure proper democratic and true consultation is seen to be done;

8. Transition to cleaner solutions to be rapid and effective.

Part 3

ERADICATING ECOCIDE

'We tolerate without rebuke the vices with which we have grown familiar.'

PUBLILIUS SYRUS, Roman author, 1st century B.C.

Chapter 5

ECOCIDE: THE 5TH CRIME AGAINST PEACE

THERE are certain principles of universal validity and application that apply to civilization as a whole. They are the principles that underpin the prohibition of certain behaviour, for example apartheid and genocide. Such abuses arose out of value systems based on a lack of regard for fellow humanity and are now universally outlawed. The rendering of such action as illegal is premised on the advancement of a higher morality that operates without caveat of qualification, a morality based on the sacredness of human life. In a world aspiring to sacredness of life, it is still necessary to identify the crimes to prevent those who fail to live by similar values. But what of the well-being of *all* life – not just that of humanity – but of all who inhabit a territory over which one has certain responsibilities?

It was the humanitarian crisis of World War II which prompted the creation of the United Nations Organization whose stated aims are to facilitate cooperation in international law, international security, economic development, social progress, human rights, and the achieving of world peace. The Charter of the United Nations (UN Charter) declared in 1945:

"We the peoples of the United Nations, determined to save succeeding generations from the scourge of war...

*to promote social progress and better standards of life in
greater freedom".*[1]

In advancement of peace, the term genocide was soon given
international legal recognition to describe the enormous deliberate
destruction of human life, such as the holocaust of World War
II. Trials were held in Nuremberg to prosecute perpetrators.
However, it took over 50 years for the creation of the International
Criminal Court (ICC) to provide a permanent international
enforcement tribunal, as set down by the provisions in the Rome
Statute and ratified in 2002.[2] Jurisdiction is limited to prosecution
of individuals of the four *'most serious crimes of concern to the
international community as a whole'*,[3] more commonly known
as the four Crimes Against Peace. They are: Genocide, Crimes
Against Humanity, War Crimes, and Crimes of Aggression.[4] Now
another type of international crime against peace has arisen: that
crime is Ecocide.

THE CRIME OF ECOCIDE

The neologism ecocide is already in use to a limited extent,
denoting large-scale destruction, in whole or in part, of ecosystems
within a given territory.[5] Ecocide is in essence the very antithesis
of life. It can be the outcome of external factors, of a *force majeure*
or an 'act of God' such as flooding or an earthquake. It can also be
the result of human intervention. Economic activity, particularly
when connected to natural resources, can be a driver of conflict.
By its very nature, ecocide leads to resource depletion, and where
there is escalation of resource depletion, war comes chasing close
behind. The capacity of ecocide to be trans-boundary and multi-
jurisdictional necessitates legislation of international scope. Where
such destruction arises out of the actions of mankind, ecocide can
be regarded as a crime against peace, against the peace of all those
who reside therein. In the event that ecocide is left to flourish, the
21st century will become a century of 'resource' wars.[6]

For the purpose of international law, I propose the following
definition for ecocide:

the extensive destruction, damage to or loss of ecosystem(s)
of a given territory, whether by human agency or by other
causes, to such an extent that peaceful enjoyment by the
inhabitants of that territory has been severely diminished.

There are two categories of ecocide: non-ascertainable and ascertainable ecocide. Non-ascertainable ecocide applies where the consequence, or potential consequence, is destruction, damage or loss to the territory *per se*, but without specific identification of cause as being that which has been created by specific human activity.

Ascertainable ecocide describes the consequence, or potential consequence, where there is destruction, damage or loss to the territory, *and* liability of the legal person(s) can be determined. The destruction of large areas of the environment and ecosystems can be caused directly or indirectly by various activities, such as nuclear testing, exploitation of resources, extractive practices, dumping of harmful chemicals, use of defoliants, emission of pollutants or war. Examples of ascertainable ecocide affecting sizeable territories include the deforestation of the Amazonian rainforest[7], the proposed expansion of the Athabasca Oil Sands in northeastern Alberta, Canada,[8] and polluted waters in many parts of the world, which account for the death of more people than all forms of violence including war.[9]

In any given example of ecocide, the extent of '*destruction*', '*damage*' or '*loss*' suffered requires analysis. Whereas '*destruction*' and '*loss*' are easy to ascertain by way of data, what constitutes '*damage*' for the purpose of establishing the crime of ecocide is more complex. Size, duration and significance of impact of damage to a territory in most instances shall be of relevance to determine whether the crime is made out. The Rome Statute sets out an extended definition of damage to the environment, specifically as a consequence of War Crimes, which provides useful assistance. Article 8(2)(b)(iv) criminalises:

widespread long-term and severe damage to the natural
environment which would be clearly excessive in relation

to the concrete and direct overall military advantage anticipated.

Change one word here: 'widespread long-term and severe damage to the natural environment which would be clearly excessive in relation to the concrete and direct overall *community* advantage anticipated', and incidents of ecocide such as the BP Gulf oil spill can begin to be properly assessed.

The wording used in this section was adopted from the 1977 United Nations Convention on the Prohibition of Military or any other Hostile Use of Environmental Modification Techniques (the Environmental Modification Convention or ENMOD). ENMOD specifies the terms *"widespread"*, *"long-lasting"* and *"severe"* as

> *"widespread"*: *encompassing an area on the scale of several hundred square kilometers;*
> *"long-lasting"*: *lasting for a period of months, or approximately a season;*
> *"severe"*: *involving serious or significant disruption or harm to human life, natural and economic resources or other assets.*

These expanded definitions, which are already embedded in international laws of war, offer an existing basis upon which the international crime of ecocide can be seated at the table of the ICC. The word 'ecocide' bestows the missing name and fuller comprehension of the crime of unlawful damage to a given environment. As a crime that is not restricted to the confines of war alone, the categorization of ecocide as a crime against peace is appropriate. Thus, for the purpose of defining ecocide '*damage*', determination as to whether the extent of damage to the environment is '*widespread, long-term and severe*' can be applied to ecocide in times of peace as well as in times of war.

ECONOMIC JUSTIFICATION

The Economics of Ecosystems and Biodiversity (TEEB) study[10] is a major international initiative analyzing the global economic benefits of biodiversity, the growing costs of biodiversity loss and ecosystem degradation and the failure to take protective measures versus the costs of effective conservation. The TEEB three year study has been led by UNEP with financial support from the European Commission, German Federal Ministry for the Environment, and the UK Department for Environment, Food and Rural Affairs. Reporting has been staggered in various stages culminating in the release of the final TEEB synthesis report at the COP10 meeting in Nagoya, Japan in October 2010. TEEB has put the conservative cost of global ecocide by the world's top firms at $2.2 trillion for 2008, a figure bigger than the national economies of all but seven countries in the world. The figure for 2009 is $4 trillion.

The TEEB report sets out comprehensively the financial justification to call for a halt to all destructive practices. Its summary valuation results are a global economic indicator of the enormity of the degradation and loss over the next 40 years if business is to continue as usual.

Losing biodiversity and ecosystems has consequences for society. TEEB has assessed the impacted 'services' to be primarily food production, water regulation and climate change resilience. When a service is viewed in isolation it can be 'costed'. In the event of Colony Collapse Disorder spreading throughout the world it is almost impossible to quantify just how much that will mean as a costing analysis, such will be the enormity of impact of CCD on numerous services. This may be one service we simply cannot afford to lose. The TEEB reports conclude that comprehensive biodiversity conservation is necessary and suggests that any costing analysis has to include various types of costs including use of land restrictions, management costs for measures such as fencing and breeding programmes, and transaction costs for implementation. It is a costing analysis premised on evaluating the financial burden to humans (i.e. the cost of conservation) to ensure upkeep of various services. Prevention of loss of services is one approach; one that is likely to be expensive and exceptionally vulnerable to extensive cost cutting and money saving measures.

The starting point has not been an analysis premised on the principle of the existence value.

The existence value is the non-use value. It does not require that utility be derived from direct use of the resource: the utility comes from the resource simply existing. Many species exist without us fully understanding their 'service value' to us, which does not diminish their value. Biota hidden in the depths of the sea may have no direct impact on our lives, but the loss of it will impact other species, and if impact is on a grand enough scale it will be a contributory factor in the breakdown of numerous ecosystems, which in time will affect our peaceful enjoyment. Every being is interconnected in the web of life and value cannot be determined simply on the premise of 'service' to humanity.

Money is a useful tool of measurement; it assists us in quantifying the enormity of the damage and destruction caused. But to understand existence value is a step further – it demands a shift in thinking, one that does not subject the world to market measurements.

Eco-colonization

The land-grabs for resource exploitation of today by international corporations are a repeat of the past colonial conquering of 'virgin land' for commercial exploitation. Colonization may be relegated by many as a subject matter of mere historical interest, but in truth colonization is very much alive. Whilst the focus has shifted from human slavery to plundering of ecological resources, the mechanics have not. As in the past, resource rich territory is distributed among the corporate interlopers, their control is registered, secured in legal title, and administered with the sole and self advancing purpose of profitable gain. This is the reality of colonization in the 21st century; it is no longer confined to the enslavement of people but enslavement of the planet. In the process, extensive damage is caused without recourse to or remedy for the well-being of either the territory or its inhabitants. The deal now, as then, is secured by long-term contract, and thus the wrongful conduct towards the inhabitants by a corporation is sanctioned and legitimised by the state.

The global reach of international corporations is such that they surpass many states' economic or territorial stature. Eco-colonization can and is happening in territories sometimes the size of nations. Such is the extent of ecosystem destruction on a global scale that similar principles and legal recognition on a par with genocide are now necessary for outlawing ecocide before further resource depletion triggers more war. It is sobering to reflect that not all colonization was effected solely by nations: the discovery and occupation of *terrae nullius*[11] or the establishment of title of inhabited territories by other means (often by force) was also initiated by charter companies such as the British East India Company. This particular company had its own army to ensure control of its resources was effectively policed. Today lobbying (and sometimes closer to ground activities) ensures effective policing to make sure power remains vested with the colonizing corporation. Thus, in reality, colonization of old differs little from the colonization of today – the heedless pursuit of profit by exploiting another's resources, wreak tragic consequences on people and planet.

Territories and their boundaries change over time, as do those who have governance of those territories. However, it is the value ascribed by those who hold the governing responsibilities of a given territory to those who inhabit those territories, which governs the conduct and actions towards the inhabitants. The inhabitants of a territory can and do fluctuate. Nevertheless, it is the habitat for all those who reside there at any given point in time. It is an environment that can be regarded as a home not solely for human occupants but also can include animals, essential minerals, water and fertile land.

RESPONSIBILITY IS A MANTLE WORN BY ALL

Article 4 of the Convention on the Prevention and Punishment of the Crime of Genocide provides that genocide is punishable as a crime irrespective of whether those committing it are *"constitutionally responsible rulers, public officials or private individuals."* This is an important tenet that has been retained by the ICC: responsibility is a mantle worn by all.

It is important to exercise enforcement and deterrence of ecocide on individuals as well as states. Unlike the primary UN judicial organ the International Court of Justice, whose main functions are to settle inter-state legal disputes and provide advisory opinions, the ICC is a permanent tribunal to prosecute individuals for crimes against peace. Thus, by incorporating ecocide as a fifth crime against peace under the Rome Statute, extensive damage to the environment is actionable against persons. It was the Nuremberg Principles that established individual responsibility under international law. The International Tribunal at Nuremberg held:

> *Crimes against international law are committed by men, not by abstract entities, and only by punishing individuals who commit such crimes can the provisions of international law be enforced.*

It is proposed that ecocide be a crime of strict liability, one without the requirement of *mens rea*. [12] The reasons are four-fold. Firstly, ecocide is a crime of consequence. It is often not the conduct itself that is in question but the consequences of the conduct. For instance, a company in the business of creating and generating energy may be at risk of committing ecocide depending upon where it procures its energy. Use of extractive practices would render the operators liable, whereas procurement from renewable sources would not. Secondly, the gravity and consequence of extensive damage and destruction to the environment justifies conviction without proof of any criminality of mind. Historically, courts had assumed that since a corporation could not have a criminal state of mind in isolation from its directors, it could only be guilty of an offence which did not include any mental element. Strict liability would therefore ensure application of international governance of corporate created ecocide. Thirdly, without absolute liability for ecocide, the legislation would be rendered largely ineffective.

The fourth reason is the rationale that strict liability places the focus on the onus of first preventing the harm, not on the blame of the accused. In the case of ecocide, as with all crimes against peace, the focus is ultimately on war prevention. Further destruction of resources will rapidly dissolve into violent conflict over allocation of resources. By creating a pre-emptive binding obligation, the crime of ecocide is focused on prevention from the outset. It creates a quasi-crime, a regulatory offence, rather than an ordinary criminal offence. The concept of fault for regulatory offences is based upon the reasonable care standard, which does not imply moral blameworthiness in the same manner as criminal fault. Thus, conviction for breach of a regulatory offence suggests nothing more than that the defendant has failed to meet a prescribed standard of care, albeit a standard of care of great exactitude. Rather than starting from a premise of punishing past wrongful conduct, regulatory measures are generally designed to prevent future harm through the enforcement of minimum standards of conduct and care. In doing so, regulatory legislation involves the shift in emphasis from the protection of individual interests to the protection of public and societal interests.

Whilst recognition of the true cost of environmental damage is beginning to be more fully considered in some environmental prosecution sentencing, fines do not ultimately provide satisfactory recourse to those affected, nor does it prevent further illegal activity. The failure to govern illegal logging of the Amazon amply demonstrates this point: fines are merely factored in by the company as an externality, to be paid if and when caught. Thus responsibility continues to be sidestepped time and again. However, where ecocide is subject to criminal prosecution, the implied breach of duty of care imposes another penalty. As an international crime it is an imprisonable offence. Additionally, as a regulatory offence, restoration is required. Restorative justice is a far more powerful legal tool than mere pecuniary justice. Imposing extensive restoration provisions ensures the duty of care is not evaded by those who have derogated their responsibilities. In this manner, the acquisition of rights over a territory ensures application of the counterbalancing responsibilities owed by those in whom the rights are vested.

The adoption of ecocide as a fifth peace crime to be governed by the ICC would compel state parties to the Rome Statute and individuals therein to abide by their international legal responsibility to prevent ecocide being wreaked under their tenureship. In doing so, the prevention of ecocide would attract the legal status of *erga omnes* (Latin: 'towards all") meaning an obligation flowing to all. Accordingly, *erga omnes* obligations are owed to the international community as a whole. When a principle, in this case the sacredness of all life, achieves the status of *erga omnes* the rest of the international community is under a mandatory duty to respect it in all circumstances in their relations with each other. An *erga omnes* obligation exists to prevent the breach of a primary crime. Ecocide would therefore be included as an example of an *erga omnes* norm, alongside piracy, genocide and crimes against humanity such as slavery and racial discrimination.

Under the Rome Statute's complementarity principle, the court is designed to complement existing national judicial systems: it can exercise its jurisdiction only when national courts are unwilling or unable to investigate or prosecute such crimes. Primary responsibility to investigate and punish crimes is therefore left to individual states. Thus many signatories to the Rome Statute (but not all) have implemented national legislation to provide for the investigation and prosecution of crimes that fall under the jurisdiction of the ICC. Hence, by implementing ecocide as a crime at international level, the pressure is immediately created for the crime to be speedily implemented at national level.

There is an additional reason for seeking international recognition of ecocide: until we have correctly identified the problem, we are unable to provide the corresponding solutions. International law evolves in response to the changing world, and is by no means a perfect beast, growing and changing direction as it expands. There is missing law - law that addresses collective responsibilities. Whilst we have laws that set in place legal duties of care for each other (e.g. to not steal, cause bodily harm, kill), expanding our cycle of care from human to human governance to human to non-human governance is the next step. Voluntary corporate governance, market trading and offset mechanisms are all completely incapable of stopping the ecocide. Creation of the crime of ecocide creates a pre-emptive obligation to act responsibly before damage or destruction of a given territory takes place. In doing so the burden shifts dramatically, sending a powerful global message to the world, of a premise that applies to us all, not just to those involved in business or waging war, to take responsibility for the well being of all life – human and non-human.

Chapter 6

THE SACRED TRUST OF CIVILIZATION

The United Nations and the Power of Language
The initial building blocks for the advancement of peace and freedom between nations, not just for a select a few but for all, were born in the 1940s. Impetus to set in place an international peacemaking organization stemmed in large part from the inability of its predecessor, the League of Nations, to prevent the outbreak of the World War II. Ten European leaders – Belgium, Czechoslovakia, France, Great Britain, Greece, Luxembourg, the Netherlands, Norway, Poland, and Yugoslavia – met in London on June 12th 1941. It was an auspicious day: the ten countries signed the Inter-Allied Declaration, vowing to work together for a free world. Two months later, on August 14th, U.S. President Franklin Delano Roosevelt (1882–1945) and British Prime Minister Winston Churchill (1874–1965) signed the Atlantic Charter, in which they outlined their aims for peace. A postwar order was to be supported by an effective international organization. During this meeting, Roosevelt privately suggested to Churchill the name of the future organization: the United Nations.

In late 1943 Roosevelt, Churchill, and Soviet premier Joseph Stalin (1879–1953) met in Teheran, Iran. They agreed on the responsibility of a United Nations organization and pledged their commitment to work together to prevent war from happening again. Cogs began to turn, meetings were held and planned in quick succession.

Franklin Roosevelt died just weeks before the inaugural conference of the United Nations. He was a man who understood the power of language to frame and shape human aspiration and a key visionary of the organization. The day before he died he wrote his last speech, in honour of Thomas Jefferson's birthday, to be given the following day. Naming the root causes of conflict as fear, ignorance and greed, he surmised:

if civilization is to survive, we must cultivate the science of human relationships – the ability of all peoples, of all kinds, to live together and work together in the same world, at peace...

In his last words he appealed to people to dedicate themselves to peace and act with faith.

The only limit to our realization of tomorrow will be our doubts of today. Let us move forward with strong and active faith.[1]

Roosevelt's legacy was the creation of an organization premised on the conviction that it was the peoples as a whole – not merely their heads of state – who could ensure peace for all. Cultivating harmonious relationships, understanding our interdependency, was key to the future of us all. Echoing the language Roosevelt deployed, the language embedded in the UN Charter was carefully chosen and deliberated on, each precise word and turn of phrase laden with the spirit, meaning and principles to shape an altruistic civilization.

Three weeks later the war with Germany ended, and Roosevelt's aspiration for a peace-making organization came into being soon after with unanimous adoption of the governing treaty, the Charter of the United Nations. On October 24th 1945, shortly after the war ended with Japan, the United Nations Organization entered into force. The UN Charter was and remains the primary defining international legal document pursuing peace premised on the respect for the principle of equal rights and self-determination of peoples in pursuit of greater freedom.

THE SACRED TRUST OF CIVILIZATION

It wasn't only the countries directly involved in World War II that were affected and required help. Other territories detached from 'enemy states' were suddenly in the spotlight. The fate of the colonies of the losing powers had to be determined. The spoils of war were to be shared amongst the succeeding powers, but this time, after the collapse of the League of Nations, it was resolved to strengthen the principle as laid down by their predecessors; that of the sacred trust of civilization. Paradoxically, the setting out of the sacred trust was to ensure that those who were ruling over other people's territories adhered to their obligations: it was a governance mechanism of member states, not of the people they were governing. Set out in Article 73, one particular responsibility has proven to be the most testing of state obligations:

> *Members of the United Nations which have or assume responsibilities for the administration of territories whose peoples have not yet attained a full measure of self-government recognize the principle that the interests of the inhabitants of these territories are paramount, and accept as a sacred trust the obligation to promote to the utmost, within the system of international peace and security established by the present Charter, the well-being of the inhabitants of these territories...* [2]

The sacred trust sets out the duty that is owed by the governing state to those on whose behalf they rule, that is a) that the interests of the inhabitants override all other considerations; and b) that they will promote to the utmost the well-being of the inhabitants of the territory.

The concept of the sacred trust is steeped in the annals of philosophical thought throughout history. One proponent was the 18th century Irish statesman and British MP, Edmund Burke.

Burke believed that the British colonies should be administered with trust for the wellbeing of local inhabitants, an idea that ran counter to the discriminatory practices of the time. Burke, speaking on the British rule of India by the British East India Company in 1783 asserted:

> *...all political power which is set over men ... being wholly artificial, and for so much a derogation from the natural equality of mankind at large, ought to be some way or other exercised ultimately for their benefit. If this is true with regard to every species of political domination... then such rights, or privileges, or whatever you choose to call them, are all, in the strictest sense, a trust.* [3]

Where political domination failed to uphold equality of those peoples, Burke reasoned, then those who govern have a duty to exercise their power for the benefit of the people over whom they have dominion. Burke argued time and again that the administrators of a colony were under a moral duty to ensure that they acted on behalf of the interests of the peoples whose land they occupied. The trading companies of that time acted from a position of neglect, commercial exploitation, discrimination, profit and general irresponsibility, without any form of accountability or responsibility for the peoples whose lands were plundered.

It wasn't until the early 20th century that the concept of a legal trust took hold as a means of placing colonial rule on an ethical, humanitarian footing to replace earlier forms of state controlled colonialism and control by corporate entities such as the British East India Company. The principle of a sacred trust was taken up once again, this time by the American President Woodrow Wilson. Wilson advocated that all colonized peoples have the right to self-determination and those who had not yet attained such status were protected in the interim by the moral and legal obligations imposed by a sacred trust of civilization. The principle was embodied in the Covenant of the League of Nations in 1919, the precursor to the United Nations. Article 22 the League of Nations reads:

To those colonies and territories which as a consequence of the late war have ceased to be under the sovereignty of the States which formerly governed them and which are inhabited by peoples not yet able to stand by themselves under the strenuous conditions of the modern world, there should be applied the principle that the well-being and development of such peoples form a **sacred trust of civilization** *and that securities for the performance of this trust should be embodied in this Covenant.*

This was the very same sacred trust that was carried forward and expanded upon in the subsequent United Nations Charter as a founding principle and legal obligation of member states. Time, however, has demonstrated that the sacred trust of civilization has all too often been neglected, its significance disregarded and, perhaps most damaging of all, its application has been constrained.

The Sacred Duty Owed to The Peoples

Chapters XI, XII and XIII (Articles 73–91) of the UN Charter set out the duty to protect the interests of self-governing inhabitants. Chapter XI applies to territories not placed under the UN trusteeship system; whereas Chapters XII and XIII address territories that have been placed under the UN Trusteeship system. The territories are termed 'non-self governing territories' (NSGTs), a more modern term for what used to be known as colonies. A central purpose of the provisions relating to non-self-governing territories is the prevention of exploitation by powerful nations.[4] With the resurgence of corporate colonization and escalation of territorial destruction, the provisions set out here are required now, more than ever, to be put to use again.

Vedanta: a Modern Day Colonialist[5]

The Dongria Kondh people are facing the modern day pillager of land. This time round, instead of it being the British East India Company it is a bauxite mining company, one of the largest in the world. Like the British East India Company this company is also

British, a FTSE-100 listed British corporation – with a difference. This time it is not a white colonialist in charge, but a fellow Indian. Mining giant Vedanta wants to destroy the Niyamgiri Hills in India and mine for the precious mineral bauxite – the base mineral used in the manufacture of aluminium – which lies underneath it on a massive scale. Vedanta has little concern for the people of the Dongria Kondh who live there. The mine will extract for 25 years and in the process will destroy their world; a territory which has existed for many, many centuries. Before the Dongria Kondh are driven out, it begs the question whether Vedanta has the right to destroy their homeland without fulfilling their duty to ensure that the well-being of the inhabitants are put first.

For the Dongria Kondh people in eastern India, the Niyamgiri Hills of Orissa in which they have lived for thousands of years are sacred. Their existence is one without typical western comforts: by UN standards, many of the children are undernourished, and less than five per cent of adults can read or write. The government of India, the state government of Orissa and the Indian subsidiary of Vedanta Resources Plc, are in agreement to dig up the Niyamgiri mountain at the rate of three million tons a year. An aluminum refinery has already been constructed at the foot of the hills. Vedanta and its supporters in the Indian government argue that this is vital for the development of the new Indian nation and will bring jobs and infrastructure to some of the poorest people on the planet. The right of the Dongria to determine their own fate was being ignored.

Orissa is a remarkably fertile land, producing an abundance of fruits and rare medicinal herbs. Their lands could supply the world with Ayurvedic medicines without destroying their territory. Many millions of people could benefit from the produce harvested from their soils, generating jobs and health for centuries to come. Sharing of the land, equitably between the inhabitants and the wider community, is not an option. Instead, the corporate right to take and despoil has taken precedence over the peoples' sacred trust. For the people of the Dongria, the law that governs them is the law of the land. To them, the Niyamgiri Mountain is better known as the Mountain of Law.

Chapter XI is of specific relevance to indigenous communities who are non-self governing and whose inhabitants' well-being has been threatened or diminished by the exploitative activities of a state, or its agents (whether it be through the activities of a nationally registered or overseas corporation), which govern their territory. The principle of a 'sacred trust' incumbent upon an administering state of a colonized territory has lost currency in recent years. The neglect is in part due to the perceived diminishing responsibilities that member states had for the old colonies. NSGT was a term that had become synonymous with colonized territories of the past, but few acknowledged colonies of today. As a result little use was made of this exceptional provision; the perception was (and continues to be) that colonization is no longer in existence.[6] With only 16 NSGTs remaining, the expressed desire is to see the remaining territories discharged as soon as possible.[7] To date, no territories have been added to the existing list since 1950, but Hans Kelson, jurist and foremost expert on the laws of the UN, argues there is nothing to prevent non-self governing territories applying to be recognized as such.[8] There is a convincing argument that an ongoing legal duty is owed by the UN, to provide remedy under Chapter XI for those now facing the modern-day reality of commercial eco-colonization, that can no longer be ignored.

The sacred trust oversees a relationship between the administrative member state, the peoples under their ward who have been subjected to (or are at risk of) colonization and who are unable to self govern, and the territory they inhabit. The principle at stake – the well-being of the inhabitants – takes precedence when an issue is to be determined with regard to the administration of the territory. Articles 73 and 74, which constitutes obligations that apply to all member states,[9] are legally binding internationally and is enforceable by way of recourse to the provisions of the UN Charter when breached. The duty *to promote to the utmost... the well-being of the inhabitants* sits at the heart of the universal principle of the sacred trust of civilization.

Under the existing obligation of a sacred trust upon member states, remedy for a breach of the trust is applicable when:

a) the people have been subjected to a decision that is not of their own making with regard to the territory;

b) it is a decision that has failed to ensure that the interests of the inhabitants remain paramount, in favour of commercial exploitation, and

c) in doing so, the member state has abrogated their obligation, to promote to the utmost the well-being of the inhabitants of the territory.

Colonization exists yet again when commercial exploitation and destruction of resources takes precedence over the obligations of the sacred trust.

It is the conscious decision to adopt certain values that brings about the civilizing (or otherwise) of humanity. Civilization is the attainment of the stage of human social development and organization that is considered most advanced. It is also a process by which a society or place reaches this stage. Each successive generation brings a deeper understanding, paving the way for responsible governance – governance in accord with a higher truth. In this way a sacred trust of civilization is understood, developed and embodied in thought and action.

A sacred trust – like any trust – has trustees, those who administer the trust. In law, a trustee's duties are based on the equitable notion of conscience and conscionable conduct. Prioritizing of personal, professional and business interests are improper; what comes first are the duties of service to ensure the well-being of all beneficiaries. The trustee acts on behalf of the interests of the beneficiary, not for his or her own interest. Therefore, for the purposes of the sacred trust of civilization, it is the member states who are the trustees bestowed with the position of responsibility and who have what is termed as a 'fiduciary duty' to the beneficiaries of the trust asset. More specifically, as the responsibility is vested in humans, it is legally known as 'superior responsibility' which rests with the appointed head of government and other senior members of the ruling state. The asset to be administered by the trustees is the territory, and the beneficiaries are the inhabitants.

The role of trusteeship adheres to general principles of equity to confer moral obligations onto the member state as the trustee. A fiduciary is the 'legal person' who has undertaken to act for and on behalf of another in circumstances which give rise to a relationship of trust and confidence. A member state, as the fiduciary, therefore owes a superior duty to protect the trust asset from unreasonable loss for the benefit of the inhabitants.

Appointing Guardians for Damaged Lands

By way of analogy, in formative years a mother owes a duty of care to her child (as does a father). Her duty is to ensure the well-being of her child, for that child is dependent on her parental care to ensure his/her very survival (it is for this reason that a governing member state of a non-self governing territory is sometimes referred to as 'the mother state'). In her role as a parent, her duty to that child as primary carer can be extended to others as more children arrive into the family circle. Motherhood is a role specific to ensuring the well-being of her child, a responsibility that diminishes as the child enters into adulthood. When she abuses her child, or fails to act to protect her child's interests, she fails to uphold her responsibilities to that child. In recognition of the child's inability to defend him/herself, laws have been put in place to provide remedy when a mothers fails in her responsibilities. In such an instance, the court will appoint a guardian to represent the child, to speak on his/her behalf and ensure his/her well-being is addressed in the course of the proceedings.

Replace the child with the planet and the mother with a corporation – for instance a logging company in the Amazon – and a very similar scenario exists. The Amazon, like the child, is unable to speak of the damage that has occurred and what it requires to ensure future well-being. Unlike the child, it has no recognized rights in law and as a consequence no responsibility is identified as being owed by those logging the territory. If caught, the company will be fined for logging unlawfully, nothing more. Without the recognition of the Amazon's rights and the corporation's responsibilities to the Amazon, a guardian cannot speak on behalf of the territory in court and individuals in the company cannot be effectively held to account. Those in the company who are vested with responsibility of the well-being of the Amazon owe an over-riding duty of care to the territory within which they are working. Where that duty of care has been breached, the fiduciaries – the directors – have failed to fulfil their moral obligation to prevent unreasonable loss, damage and destruction.

Under Article 73 of Chapter XI, a sacred trust arises when a group of people have been or are subjected to colonization and are unable to self-determine. Just as slavery is a breach of the sacredness of life, so too is colonization. Both are infringements, yet all too often the veil of corporate activity hides the injustices taking place to both the people and their territories. Close interpretation of Article 73 however takes one step further on from the adherence to the universal principle of inherent value of human life. Article 73 specifically addresses the inherent value of all life. Member state control is thus restrained by the sacred trust obligation placed upon them to prevent such conduct impinging upon the well-being of not just the peoples who live in the territory but *the inhabitants.*

A sacred trust is a state of doing which, under Article 73 of the UN Charter, applies in certain circumstances. It is an obligation that is vested in the state when a people are unable to self-determine. Whosoever has the responsibility for the determination of a given territory has the obligation to promote to the utmost the wellbeing of the inhabitants. By way of a parallel situation in a child care case (where the parent is unable to govern), when determination is being made with regard to the upbringing of the child or the administration of a child's property, the child's welfare is the court's *"paramount consideration"*. Likewise, in the absence of a people being able to determine any question with respect to the administration of a given territory, the inhabitants' well-being is the paramount consideration of the member state.

Thus a sacred trust is both a state of being and of doing, which when adhered to prevents the abrogation of the universal value of the sacredness of life. The very use of the word 'sacred' underlines the importance such a trust is accorded, reinforcing the moral as well as the legal obligation it imposes on all, not just those states governing colonized territories. Articles 73 and 74, as Hans Kelson the foremost jurist of the 20th century noted, constitute obligations that apply to all member states. [10] This is expressly stipulated with respect to the provisions of Article 73, *"Members…accept … the obligation to…"*, an interpretation that has been upheld and referred to in various resolutions adopted by the General Assembly. [11]

A PROTECTED RIGHT: THE PEOPLES' RIGHT TO SELF-DETERMINATION

The recognition of the right to self-determination by peoples whose lands had been appropriated was, like the sacred trust, in itself not new. Spaniard Francisco de Victoria, Professor of Sacred Theology in the University of Salamanca, advocated the importance of honouring the right of the indigenous Indians of Latin America to decide on their community issues as far back as the 16th century. His voice of conscience and reason likewise identified the correlating positive duty of a ruling elite to shoulder their responsibilities as guardians, rather than merely disregarding those who were perceived as primitive minorities and therefore of lesser value. Such an ethical imperative was premised on the recognition of being in service to the welfare of fellow humans regardless of skin colour and cultural practices. But it was the very 'otherness' of an indigenous community, so visually identifiable as a result of skin pigmentation, which was frequently used to justify discriminatory behaviour in the pursuit of supremacy of ownership. To accept the concept of self-determination inevitably involved the tacit acceptance of equality, a proposition that was the very antithesis of colonial rule. Whilst such lofty sentiments were deemed admirable by some, and were subsequently expanded upon by philosophers and thinkers, they remained closeted within the cloistered confines of western theological thought for another 400 years despite Edmund Burke's frequent entreaties.

But, like all truths that are hidden away in closets, there comes a time for their re-emergence. The League of Nations proved to be ultimately an unsuccessful outing, but one that helped pave the way. Twenty-six years later the United Nations Organization set down its anchor. This time round, the principle of a sacred trust was securely embedded. Article 55 of the UN Charter set down the explicit obligation to create

> *conditions of stability and well-being which are necessary for peaceful and friendly relations among nations based on respect for the principle of equal rights and self-determination of peoples.*

Self-determination was formally recognized as the autonomy of an affected people to determine their fate (which may or may not result in eventual creation of independent states). Colonialism with all its attendant discriminatory behaviour was declared a thing of the past, not to be repeated. Every nation that signed the UN Charter was obligated to ensure self-determination of their wards. Territories that had been in the clutches of others were to be released from their shackles to move toward eventual self-governance. Chapters XI, XII and XIII set out the mechanisms to facilitate the governance of affected peoples of colonized territories.

The principle of equality and self-determination of peoples laid down "two complementary parts of one standard of conduct"[12] upon which successful relations are to be built, declared the UN Charter. Under Article 1(2), one of the four purposes of the UN is:

> *To develop friendly relations among nations based on respect for the principle of equal rights and self-determination of peoples, and to take other appropriate measures to strengthen universal peace.*

Crucial to the drafting of the Charter was the explicit and repeated use of the word *'peoples'* – not nations, nor states. Here was a document that purports to be written by the people for the people, as embodied in its very first words in Article 1: *'We the peoples'.* It did not, very deliberately, commence with the words 'We the heads of nations' or 'We the heads of state.' To understand the thinking behind the drafting of the UN Charter, statements from the time shed important light on what was understood to be the differing definitions of 'nations' and 'peoples'. One in particular stands out; the 1945 Memorandum prepared by the UN Secretariat in response to a proposal to replace the word people with state. The document clarifies the intention of the UN Charter to extend equal rights to states, nations and *"peoples".*

> *The word 'nation' is broad and general enough to include colonies, mandates, protectorates, and quasi-states as well as states... 'nations' is used in the sense of all political entities, states and non-states, whereas 'peoples' refers to groups of human beings who may, or may not, comprise states or nations.* [13]

Peoples, it is clear, for the purpose of the UN Charter was defined as an all-embracing term, neither limited or restricted in application. Peoples, as well as states and nations, are all equal in the eyes of the Charter, the principles of equality and self-determination being contingent upon each other to be realized fully. Quite simply, one without the other would render the principle empty. Where there is equality for the people, there also exists the right to self-determine. Withdraw the right to self-determine and all equality between peoples is lost. Peoples can be a tiny community or a nation state – the application of the principle remains the same for all. Withdraw it and in its place inequality and autocracy foster discontent and conflict.

The principle is anchored in another well known self-evident truth, one which was espoused in the American Declaration of Independence of July 4th 1776: *'all men are created equal.'* Taking this to mean both men and women, the Charter nevertheless ensures that it remains gender balanced, reaffirming *'the equal rights of men and women.'* Thomas Jefferson's phrase in the

Declaration of Independence countered the notion of the divine right of Kings. Likewise the phrase *'respect for the principle of equal rights and self-determination of peoples'* is a rebuttal to the notion of supremacy of colonial rule. Colonial domination was no longer to be tolerated. Instead it should have opened the door to indigenous and minority peoples being given an equal seat at the table of decision-making.

THE MYTH OF SALT-WATER AND SKIN PIGMENTATION

Self-determination applies to us all. Attempts have, however, been made to restrict use by people or communities which have been perceived to be a threat to national sovereignty. Nevertheless, it cannot be disputed that the right to self-determination is understood to be a fundamental principle of international law from which there shall be no derogation. Despite being signatories to the core UN Charter objective *'To develop friendly relations among nations based on respect for the principle of equal rights and self-determination of peoples, and to take other appropriate measures to strengthen universal peace'*, many states have clung, and still try to do so, to the concept of sovereignty overriding the applicability of the right of self-determination. Not merely an aspiration, but a legally binding peremptory norm, self-determination has been reinforced by the International Covenant on Civil and Political Rights (ICCPR) and the International Covenant on Economic, Social and Cultural Rights (ICESCR) which both read:

> *All peoples have the right of self-determination. By virtue of that right they freely determine their political status and freely pursue their economic, social and cultural development.* [14]

Be that as it may, the freedom of the people of a given territory to determine their own governance without undue influence from any country has been sorely tested in recent times. [15] The path to equality and self-determination has proven to be a steep one to climb, fettered as it is by twists and turns, with many road-blocks to be dismantled. Minority and indigenous rights have time and again been marginalised at national as well as international level, violations of the very principles that were to ensure peace. One contributing factor, often overlooked or relegated to a footnote in the evolution of the UN, is the fallacious doctrine of 'salt-water and skin pigmentation.'[16] This flawed doctrine, more than any other, single-handedly set in motion a chain of events that has not only hindered the advancement of recognition of equality but silenced the voice of millions of indigenous peoples yet again.

The term is a description of the restrictions that were subsequently introduced to restrict claims of self-determination only to those peoples separated geographically by being from 'overseas' or ethnically distinct in appearance. Laced with the overtones of racial discrimination and otherness, two UN Decolonization Resolutions, Res. 1514 (XV) and Res. 1541 (XV), in direct contradiction of the very legal foundation of the Charter were adopted by the General Assembly in 1960.

The salt-water doctrine is set out at Principle IV of Resolution 1541:

Prima facie there is an obligation to transmit information in respect of a territory which is geographically separate and is distinct ethnically and/or culturally from the country administering it.

The adoption of the two Resolutions contravened the very essence of the intent, law and spirit of the Charter. In their haste to see the remaining colonial territories as independent and to bring colonialism to a speedy and unconditional end, restrictions on future application were imposed without full cognizance of the implications. In an ironic twist of fate, by confining a colonized territory to a select group who fitted criteria of little validity, a subconscious conviction in colonialism applying only to others elsewhere was exposed.

The concept of a colonized state was from 1960 suddenly stamped with a specific identity – and the stereotype was born.

Although Resolution 1541, with its damning stamp of salt-water and skin pigmentation doctrine, applied only to Article 73(e) (how to determine whether an obligation exists to transmit information about a colonized territory) not to the Chapter itself, it nevertheless became the accepted, albeit restrictive and racist, premise for ensuring that the universal right of self-determination could now apply only to a very narrow category of peoples. Despite Resolutions being considered non-binding [17] and therefore legally unenforceable, the salt-water doctrine to determine applicability rapidly became the accepted norm. Two years later, with the desire to hasten the demise of colonization, and the door now closing on future applications, the Special Committee on Decolonization was set up. Colonized territories, it was determined, were 'over there, of other peoples and of another time.'

In one fell swoop, indigenous peoples inhabiting territory within a metropolitan state who sought reliance on Chapter XI were excluded. However, despite the salt-water doctrine attempts to restrict the exercise of self-determination to the dismantling of European colonial empires abroad, subsequent developments have supported the flawed principle. The obligation imposed on member states to promote the well-being of the inhabitants of non-self-governing territories, as well as the two general statements of the principle of self-determination in Art.1(2) and 55, are not justifiably qualified by the language of Res. 1541. As the era of UN Decolonization draws to an end, it is evident this right is by no means of limited historical application or a spent force. The national self-determination of Bangladesh proved successful in the early 1970s despite lack of salt-water, and several judicial opinions by the UN's International Court of Justice have pronounced unequivocally that self-determination of the peoples is a right that trumps state sovereignty. As eloquently stated by Dillard J. in his advisory opinion on the 1975 *Western Sahara* case:

> *It seemed hardly necessary to make more explicit the cardinal restraint which the legal right of self-determination imposes (upon the governing state). The restraint may be captured in*

one sentence. It is for the people to determine the destiny of
the territory and not the territory of the people...[18]

The right to self-determination, proclaimed the ICJ, is the freedom of the people of a given territory or national grouping to determine their own outcomes and how they will be governed without undue influence from any other. More recently, in the 1995 *East Timor (Portugal v Australia)*[19] case, the court took a further step forward in pronouncing that

> *Portugal's assertion that the right of the peoples to self-determination, as it evolved from the Charter and from United Nations practice, has an erga omnes character, is irreproachable.*

The Court emphasised the right of peoples to self-determine was *'one of the essential principles of contemporary international law.'* In other words, the right to self-determine, in the eyes of the law is a right that flows to all.

REFORMING THE UNITED NATIONS

It is argued here that territories that are at risk or subjected to destruction due to ascertainable ecocide (where the governing state fails to address the issue and where the ICC is unable to act, having not yet adopted the international crime of Ecocide), can seek redress by another avenue. Namely, that the peoples of a given territory be allowed to apply under Chapter XI of the United Nations for assistance and protection of their territory.

In determining whether a territory has been colonized for modern day purposes, it is proposed that it need not be an overseas entity. Although most eco-colonization continues to be perpetrated by global north companies whose corporate activities are largely rooted in the global south, national based corporations can and are also colonizing home state or nearby territories. The focus on overseas territories are in part merely a historical reflection of the politics of expansion and invasion of perceived lesser peoples and their territories through the 16th to 20th centuries. The over-riding objective is to prohibit the colonialist conduct on territory where the well-being of its inhabitants and indigenous peoples has been ignored in the pursuit of profit. Separation can simply denote that a territory is recognized to be set apart in some manner; that can include the identification of a specific territory geographically as well as ethnically and/or culturally within the confines of a member state. [20]

If the UN is to hold true to the obligations as set out in Chapter XI, the peoples who allege a breach the peoples who allege a breach have the entitlement to call on the UN to seek enforcement of the sacred trust. [21] To do so is a right that shifts to the peoples where the governing member state has abrogated their obligations and thus is in breach of their rights and benefits resulting from their UN membership. Where inhabitants suffer wrongful conduct by the actions of corporations or other agencies, and the conduct is sanctioned by the governing state, restitution cannot be confined to national jurisdiction. To confine national jurisdiction would authorize the overriding authority to be vested in those who sanctioned the colonization of their peoples' territory in the first instance.

Proposal to Reinstate the UN Trusteeship Council

Article 75, International Trusteeship System:
The United Nations shall establish under its authority an international trusteeship system for the administration and supervision of such territories as may be placed thereunder by subsequent individual agreements. These territories are hereinafter referred to as trust territories.

One of the six principal organs upon which the United Nations was founded was the Trusteeship Council. It was formed in 1945 to oversee the decolonization of those dependent territories that were to be placed under the international trusteeship system created by the United Nations Charter as a successor to the League of Nations mandate system.

Under Article 77c a member state, on behalf of the people of the territory they were administering, could apply directly to the UN for assistance. Trust territories identified in the post-war period comprised of the remaining League of Nations mandated territories and a number of former colonies still requiring assistance on their route to independence. In 1945, the Trusteeship Council was assigned eleven territories, former colonies that required assistance transitioning to independence. Collectively, they accounted for over a third of the world landmass. The Trusteeship Council operations were eventually suspended in 1994 after the last territory was granted independence.[22] As there are at present no trust territories, the Trusteeship Council is currently in abeyance for want of a role.

It is proposed that the Trusteeship Council be put back to use for the territories of peoples that have been subject to or are at risk of ecocide. Where a breach of Chapter XI has been found, as an alternative to sanctioning a member state under Article 6, the General Assembly can make recommendations to the members "on any such questions or matters" that come "within the scope of the Charter or relating to the powers and functions of any organs provided for in the present Charter".[23] It is therefore within the remit of the General Assembly to recommend, either at their own behest or by the petition of a people, that the territory in question become a recognized territory under the auspices of the Trusteeship Council of Chapter XII.

In a situation where there has been no breach of Chapter XI, a member state could, it is submitted, make use of the provisions under Chapter XII where a territory under their auspices has suffered or is at risk of non-ascertainable ecocide. Restoration of the Trusteeship Council's central role, to assist territories in need, could moreover address the pending wave of territories and peoples at risk of non-ascertainable ecocide, such as those facing the pending rising sea-levels (the Maldives, Tuvalu, Kiribati etc). A member state could thereby seek reliance on the collective responsibility of all member states to ensure the well-being (and re-housing on other territory) of their soon to be dispossessed inhabitants. In doing so, ownership would not be the determining issue, and allegations between nations of unjust enrichment disappear. Instead, the embers of a far worthier goal begin to glow: the first steps towards trusteeship of the planet to advance governance in pursuit of the well-being of all.

THE UNIQUENESS OF THE UN CHARTER

We the peoples of the United Nations determined … to promote social progress and better standards of life in larger freedom. [24]

The UN Charter is an exceptional and unusual document of international legal weight. Where most internationally, legally binding tracts concentrate on behaviour that is forbidden (and thus actions that are unlawful), the UN Charter legislates upon the principles and behaviour to be adhered to (principles that are imposed as legal duties and obligations), to advance peace for all humankind. The very remit of the UN is predicated on that which ought to be done, not merely that which is to be forbidden. It is setting the bar of good standards to which, when applied and built upon, result in peace. The Charter is more than an aspirational document, however. It was and remains a constituent treaty, binding upon all member states. To that end, the Charter states that obligations to the United Nations prevail over all other treaty obligations. In other words it is the principal legal document of global governance. It has not always been interpreted as such.

The purpose of the preamble is to serve as an interpretative guide for the provisions of the Charter through the highlighting of some of the core motives of the founders of the organization. The promotion of '*social progress and better standards of life in larger freedom*' is a valuable principle – one that nations through the centuries have espoused. But somewhere along the line in the last 65 years since its inception we seem to have created a world drowning in a myriad of rules, regulations, laws and legislation based on fear and ownership. *'Better standards of life in larger freedom'* seems to have faded into a small point on the horizon of our ever more cluttered yet increasingly disconnected empirical world. Social progress seems to have been displaced by social regress.

The UN Charter provides a premise upon which to build a new platform based on peace and trusteeship. The maintenance of *'justice'*, of moral rightness based on ethics, law, natural law, rationality and fairness, must surely embrace ecological justice. *'To unite our strength to maintain international peace and security'* are ultimately dependent on functioning ecosystems. To ensure ecological justice is to ensure that ecosystems remain intact and functioning. That includes *'the peoples.'*

Chapter 7

HOLDING
BUSINESS TO
ACCOUNT

Taking responsibility for the environment is a relatively new arena in the evolution of identifying duties and obligations. World War II was a turning point in the evolution of identifying our responsibilities to our fellow humans. The Vietnam War proved to be a moment in history that caused us to question our responsibilities to all life, not just human life. It raised questions, but failed to impose responsibilities.

There is no single treaty which establishes principles and rules of general application to all hazardous substances or activities. The international community has, however, adopted a broad policy guideline; back in 1972 Principle 6 of the Stockholm Declaration stated that the

> discharge of toxic substances or of other substances and the release of heat, in such quantities or concentrations as to exceed the capacity of the environment to render them harmless, must be halted in order to ensure that serious or irreversible damage is not inflicted upon ecosystems.

Over 40 years on and Principle 6 of the Stockholm Declaration has still yet to be made law.

Turning a Blind Eye

Preparations for the 1992 Rio de Janeiro Earth Summit were extensive. It was to put in place the Climate Change Convention which in turn would lead to the Kyoto Protocol. The fourth United Nations Conference on Environment and Development preparatory committee meeting on Climate Change was to be held in advance in New York over four days. It was to be an auspicious meeting, the culmination of a three year negotiation process which would lay the foundations for all subsequent climate change negotiations. Sweden and Norway had drawn up a document called Agenda 21 – the agenda for the 21st century – containing detailed proposals for legally controlling multinational corporations. Major stakeholders to the Agenda 21 proposals were notified of the impending proposals and invited to participate.

In an ironic twist of fate, it was not those directly impacted by corporate activity who were invited to participate (NGOs had 'consultative status', the granting of 'observer status' to indigenous community organizations came years later – both terms that ensured that public and environmental concerns were kept at arms length); instead the participants were the very industry players whose activities had been identified as the problem. Those invited were a selection of the world's top corporations. They were the stakeholders (a new word coined by R Freeman in his book *Strategic Management: A Stakeholder Approach* in 1984) who were given participatory status. It did not take long for them to organize themselves into two groups, the BCSD (Business Council for Sustainable Development, comprising 48 business leaders from major corporations all over the world) and the ICC (International Chamber of Commerce). The majority of both group members were also members of the Global Climate Coalition, an industry lobby group set up by the American Petroleum Institute (API) in 1989 to lobby against anything that might adversely affect the oil industry when it came to climate change. The Global Climate Coalition was active from the outset.

BCSD presented a high profile media campaign publicising its commitment to full environmental cost accounting, with bold rhetoric centred on the 'changing course' of industry. Meanwhile the ICC, who refuted all claims that climate change was anthropogenic, lobbied to remove all cost accounting measures. Nevertheless, despite the stated polar positions of the two groups, over half of the BCSD companies were on the board of the ICC. All three groups had an enormous say at the outcome of the preparatory meetings held in New York. On June 14th 1992 at the Earth Summit the full text of Agenda 21 was revealed and 178 governments voted to adopt the program. Agenda 21 recommended that companies should demonstrate a commitment, in respect of toxic chemicals, *'to adopt standards of operation equivalent to or not less stringent than those existing in the country of origin.'* The final draft omitted all proposals for controlling multinational corporations. All mandatory provisions had been removed and instead were replaced with a commentary on the importance of business and industry to ensure environmental protection. Voluntary measures that corporations were already taking were highlighted, references were made to eco-efficiency and the precautionary principle was presented as the appropriate course of action. Corporations like Shell and Du Pont, both members of the Coalition and the BCSD, spoke in press conferences of their *'precautionary approach'* to global warming.

After the BCSD press conference in Rio, Italian oil giant ENI President Gabriele Cagliari was asked if the world can burn all the oil on the planet and call it sustainable. Cagliari, seemingly unaware of the incongruity of the situation, simply answered *'Yes.'*[1]

The Rio Earth Summit revealed a fundamental flaw: the majority stakeholders in the issue had not been properly identified and given a voice. It was a summit that sought to confront the root causes of the climate crisis affecting the planet. Instead, it became a meeting for corporations to stamp their authority and their supremacy over governmental decision making. No recognition was given to the stakeholder with most to lose – the planet. No platform was provided for the communities most adversely affected by climate change and the destructive activities of the main extractive businesses.

Agenda 21 comprehensively failed to regulate damaging corporate activity. Chapter 30 paragraph 3 advocated

> *Business and industry, including transnational corporations, should recognize environmental management as among the highest corporate priorities and as a key determinant to sustainable development.*

In a similar vein, chapter 35 paragraph 3 pronounced

> *In the face of threats of irreversible environmental damage, lack of full scientific understanding should not be an excuse for postponing actions which are justified in their own right. The precautionary approach could provide a basis for policies relating to complex systems that are not yet fully understood and whose consequences of disturbances cannot yet be predicted.*

The words *'could'* and *'should'* connote a suggested route of action, a possible voluntary policy at best, without mandatory compliance. At its highest, the burden of responsibility was placed as a recommendation, but nothing more. The Earth Summit was lobbying at its worst. Meetings had taken place between government officials and corporate lobbyists in private rooms, hidden and secretive. Transparency of process and inclusion of other relevant parties in discussions and decision-making did not feature as part of the Summit process. Sadly not much has changed in the past 18 years since Rio; climate negotiations continue today on the same unbalanced premise. Each year, world leaders fail to resolve the problem and history continues to repeat itself until we learn to re-evaluate and change our approach.

The 1992 Rio Earth Summit was a compromise on a grand scale for the planet. Governance of environmentally damaging corporate activity was rendered toothless by the final documents which contained only voluntary statements and unfulfilled promises. Principles were set out, but with no binding legal mechanisms to ensure they were enforced. Such strong words could only be considered to be an aspiration at best. Principle 13 is one example:

> *States shall develop national law regarding liability and compensation for the victims of pollution and other environmental damage; they shall also co-operate in an expeditious and more determined manner to develop further international law regarding liability and compensation for adverse effects of environmental damage caused by activities within their jurisdiction or control to areas beyond their jurisdiction.*

Eighteen years on from the Earth Summit and there is still no legal definition in force under international law as to what constitutes a hazardous activity or substance, and many industrial and other activities which may, over time, pose significant long-term environmental threats are not subject to specific international environmental regulation. Dedicated international rules are virtually nonexistent for transport, mining, agriculture and energy. The rules that do exist have not been developed or applied in the framework of co-ordinated regulatory strategy. Globally harmonised rules establishing high standards of human and environmental protection are necessary but do not yet exist. Although industrialized countries have put in place a complex body of binding legal obligations under regional agreements, EC law and OECD acts, the extent to which many of these rules are adhered to in developing countries is questionable and difficult to monitor.[2]

CONVENTION ON CIVIL LIABILITY FOR DAMAGE RESULTING FROM ACTIVITIES DANGEROUS TO THE ENVIRONMENT, 1993

There is one light that hasn't been entirely extinguished however. One year after the Earth Summit, Principle 13 was very nearly realized. A European law was put on the table setting out liability and compensation for adverse effects of environmental damage, as had been specifically called for. The law is a European Convention: after the outcome of the Earth Summit it was opened for signatories to sign. Despite the commitment voiced under Principle 13 not one country signed it. Named the Convention on Civil Liability for Damage Resulting from Activities Dangerous to the Environment, it is law which sets out definitions for dangerous substance, environment and incident. It imposes strict liability upon corporations who engage in dangerous activities, such as deep sea drilling. All that this requires to make it law is for it to be ratified by EU countries. Non-EU countries, such as America, require a similar law. It is a simple step to take; a tangible solution which fulfils the avowed commitment stated in Principle 13. Governments would finally put in place a law that specifically addresses the responsibilities of those who undertake activities that carry a risk of extensive damage and destruction on an enormous scale to the environment.

Article 2 (2)(a) defines *'dangerous substance'* as:
substances or preparations which have properties which constitute a significant risk for man, the environment or property. A substance or preparation which is explosive, oxidizing, extremely flammable, highly flammable, flammable, very toxic, toxic, harmful, corrosive, irritant, sensitizing, carcinogenic, mutagenic, toxic for reproduction or dangerous for the environment.

Article 2(10) *'Environment'* includes:
- *natural resources both abiotic and biotic, such as air, water, soil, fauna and flora and the interaction between the same factors;*
- *property which forms part of the cultural heritage; and*
- *the characteristic aspects of the landscape.*

Article 2(11):
'Incident' means *any sudden occurrence or continuous occurrence or any series of occurrences having the same*

origin, which causes damage or creates a grave and imminent threat of causing damage.

The Convention applies to all incidents causing damage or threat of imminent damage (defined extensively so as to include injury to life, property and the environment (Article 2(7)). The central provision deems an operator of an installation strictly liable for damage caused by engaging in a dangerous activity (Article 4(1)). In determining the causal link between an incident and damage, courts are required to take into account the inherent danger in the activity that such damage will occur (Article 10) and parties are required to ensure that operators either participate in a financial security scheme or have a financial guarantee to cover their liability under the Convention (Article 12). Individuals are entitled to access information relating to the environment held by public authorities or bodies with public responsibilities for the environment (Articles 14 and 15) and environmental organizations are permitted to request at any time the prohibition of dangerous activities or an order requiring that the operator take measures to prevent an incident, or take measures of reinstatement (Article 18).

To put it in context, the Convention could very easily have provided a legal route for the current BP Gulf oil spill crisis. Governments have clearly stated in Stockholm, Rio, and numerous other Summits their concern for the environment and a collective desire to do something about it. Bold words yes, but political will has yet to be backed by hard law. Here is just such a law to hold to account operators of installations for damage caused by dangerous activities. Existing laws to address the Gulf oil spill crisis are very clearly insufficient; it takes a disaster of this magnitude to alert us to this fact. Deep sea drilling clearly carries with it a risk of destruction of such enormity that corporate responsibility cannot be evaded any longer.

TRANSPARENCY OF PROCESS

Lobbying as a legitimate, transparent and public exercise in rational persuasion and information provision has a role in all political decision-making, a process used by all who seek to influence issues of importance. Where lobbying has departed from open discourse, with persuasion taking place behind closed doors, the likelihood of partial representation is inevitable. Without public accountability and transparency of any lobbying process, where wider stakeholders have the opportunity to give voice to their concerns, a balanced and reasoned decision is unlikely to be determined. A lobbyist's job is in some respects akin to a court advocate; he or she has an argument to present and has at his or her disposal various tools to assist – primary evidence, witnesses, justification – in the pursuit of exerting influence. Instead of a judge, a government official or those in a position of power determine the outcome. In a court of law, before any decision is made, the respondent has equal opportunity to make representation, and where the litigant is in person rather than being represented by a lawyer (either by choice or necessity) that person is given extra time and assistance to ensure their concerns are given fullest consideration before judgment is determined. It would be inequitable to proceed to a hearing with only one side attending. Civil procedure rules exist to ensure that proceedings commence without a respondent only in limited circumstances, for instance where the application is not disputed. As all court proceedings are held in public (save exceptional proceedings involving national security), the process involved is in effect open to daily public scrutiny. Lobbying has no rule-book, and no Code of Conduct. Those who are lobbied are rarely trained in the art of exercising independent and impartial judgment.

Lobbying cannot continue in this manner without some form of checks and balances. The imbalance is now so disparate that those who seek to raise issues from a non-industry perspective are drowned out by the sheer financial wherewithal of corporations who are capable of mounting offensives on a huge scale. Accountability of practices are urgently required, without which the norm will not change. The starting point is implementation of independent scrutiny and investigation of those who refuse to provide comprehensive breakdown of expenditure and activities undertaken. All companies with business activities that impinge upon human and environmental concerns, and which are unanswerable to the public are evading their responsibilities to the wider community of all who are affected by their activities.

The nub of the problem is not lobbying per se but the abuse of the practice. In this context, the evasion of responsibility and any form of scrutiny of practices results in lobbying purely for the benefit of business. Because business is driven by the primary consideration of profit above all other considerations, companies will only lobby for the type of regulation that makes them more competitive. Considerations for social and environmental concerns are secondary. Consequently, corporate lobbying focuses on what profitable gains are to be had from the imposition of further regulation. Corporate lobbying does not work from the premise of caring for the wellbeing of the community.

As a rule of thumb, he who has the greater potential to cause damage carries the burden of superior responsibility. Equally those businesses whose activities can cause damage, destruction to or loss of life owe a superior duty of care. Whilst this obligation remains unrecognized by most business people who are in the driving seat, so to say, this is one vehicle that continues to career out of control. Failure to protect the interests of the general public and the wider environment demonstrates the destructive influence of lobbying which has been allowed to ignore its public and democratic responsibilities.

Regulated Lobbying in the City & County of San Francisco, California, USA

The San Francisco Lobbyist Ordinance is a comprehensive law that regulates lobbyists who attempt to influence City officers on local legislative or administrative action on behalf of private parties. The purpose of the law's public disclosure requirements is to protect the public confidence in the government and to reveal lobbyists' efforts upon decision-making entities.

Key elements of the law include detailed registration requirements identifying the nature of lobbying services; quarterly reporting requirements include mandatory disclosure of payment received or promised, itemised expenditures, and political contributions made to officials; and the giving of gifts to any government officer is prohibited over a value of $50.

According to the City & County of San Francisco Ethics Commission, the agency charged with enforcing and administering the Ordinance, the payment disclosure requirements promote transparency and provide the public with specific information about the influence of private interests upon government officials and entities. In addition, all reports and registration forms are published online.

Lobbying is all too often imbalanced. At worst, rather than issues being aired in a public forum, the lobbyist is seated in the privacy of the equivalent of the judges' chambers, with the respondent left outside the court doors or, having not even been notified of the hearing, not even in attendance. There is no presumption in favour of the process being balanced and fair, and where it fails to be so, there is rarely a route to appeal.

Most corporate lobbying is closeted, secretive, and partisan. Accountability of commercial lobbying activity is limited at best. Likewise there is a lack of governance of those who are lobbied. No register exists for the public to scrutinise, no advance listing of company, lobbying issue and MP attended, nor a transcript or recording of any given meeting. All of the above is available to the public for court hearings, but no such system exists for accessing information about lobbying of our democratically elected members of parliament. As a representative of the electorate, he or she is elected by the people to represent the interests of his or her constituents. In the UK, where members of parliament account for their expenses in a public register, real-time electronic online diaries of constituency-related work are viable and easy to implement. Such a system would go some way to addressing the inequality that exists between privately (and or corporate) funded professional lobbying and community response to public issues.

In the political arena, the facility to hear and accommodate contrary opinions from the various stakeholders in a fair and democratic manner is essential for good communication. Where lobbying ceases to be concerned with persuading decision makers and becomes a propaganda exercise, distortion of the factual basis is frequently driven by vested commercial interest. Without the opportunity for other parties to voice their concerns, an inaccurate understanding results. All too often it is the largest stakeholder who suffers. To be accurate, collectively the various groups and communities whose lives may be affected by a commercial activity are the ones who are disadvantaged by the process. Climate negotiations are a potent example of the injustice at play where corporate lobbying has an enormous say and an enormous presence, but those who are most disadvantaged have no voice whatsoever. Official standing for indigenous communities, NGOs and charities is relegated to 'observer status' only. Without an accredited organization a concerned individual cannot even attend on behalf of his community and without financial support, private meetings or promotional events are out of reach of the majority. For an issue that counts as the largest public concern, the concerned public are being kept literally outside the closed doors.

Seeking to influence the outcome of a decision is rarely viewed in the light of what is for the best – all too often vested interest by those who can pay the most brings the biggest wins. Where lobbying at a political level is undertaken by corporate entities, there exists no overriding duty to ensure the interests of others are not compromised. Thus for those determined to hide their activities (or more specifically the potential adverse consequences of their activities), accountability, transparency and scrutiny are three words – which if enforced through law – would radically change the landscape of political lobbying overnight.

None of the three conditions have been in evidence in current climate negotiations, and very little is to be found in national and international politics generally. Again the scales of justice are out of kilter. Provision for the public voice has been marginalised and as a result considered to be inconsequential. Those who speak as voices on behalf of the Earth do not have a seat at the table despite their knowledge that can be brought to bear on the proceedings. At its heart lobbying is about presenting a particular unpartisan perspective of an interested group or party. It is an exercise in rational assessment unburdened by self-interest and partisan perspective. Lobbying is part of a public discourse. To deny proper stakeholder engagement is to prevent the people from being able to participate. Politics then remains firmly in the control of the few who are blinkered by corporate and economic agendas.

Governments, driven by the obsessive pursuit of economic gain, often undervalue subsequent ecological losses that can arise out of profit-making activity. Promotion of economic policies can inadvertently endorse damaging or destructive practices for the sake of immediate financial benefits, which can in turn create a mounting ecological deficit. All too often it is a deficit that is identified long after the event. Myopic financial policy takes pre-eminence over longer-term damage and destruction; by keeping the focus firmly set on the short-term, problems mount for others to address at some indeterminate later date.

A commercial entity seeking to persuade a government of a given proposition is equally unlikely to give proper due consideration to lasting environmental impact. Where law does not exist to prohibit damage, loss or destruction, corporate activity that causes ecocide will inevitably continue unabated. Evasion of environmental responsibility is ultimately dictated by financial considerations and where conflict between business and environmental concerns arises, the voices of concern are all too often drowned out by the even louder voice of profit.

Rights and responsibilities govern our human behaviour. In life, we have rights bestowed upon us at birth (we talk of children's rights); over time our responsibilities evolve as our lives enter into maturity. Prior to reaching adulthood, it is the parent who takes responsibility for the child, in whole until 10 and in part until 18, until the child has reached the age of full legal responsibility. This is reflected in law: in the UK we cannot prosecute children under the age of 10 and more serious sanctions apply after a youth attains majority at the age of 18. In adulthood, responsibility then shifts: the individual becomes responsible for his/her actions henceforth. Maturity informs our being and our ability to engage with society with equanimity, lightly shouldering our responsibilities. It is the exercising of rights by an adult and the adherence to the corresponding responsibilities that balances our behaviour. All actions have consequences: without consideration or application of our responsibilities, damage and destruction can occur. Failure to uphold our responsibilities results in the failure of humanity to live harmoniously. Thus, an adult who steals has failed to accept his responsibility to society as a whole, and although it is a minor act in isolation, if left unchecked it opens the door to theft by many others. In time theft would attain the semblance of being an accepted norm, and despite the moral shortcomings of such activity it would be accepted by the majority. The thief could exercise his/her (silent) right to steal, without the imposition of obligation to those from whom he/she had stolen. As a result, society's welfare is in turn prejudiced in favour of the unspoken protection of the individual's right to steal. No responsibility is taken for the breakdown of society until someone calls for a law against theft, thereby imposing individual responsibility for the crime of theft. It is by such legal governance that we police those who fail to uphold their responsibilities.

Children have various rights but few responsibilities. Of the few that are imposed at a young age, when a responsibility is breached the child is sanctioned by parents rather than criminal courts. As a baby, the child bears no responsibility nor is he or she accountable for his or her actions. After all, a baby is incapable of understanding such concepts. A corporation, by comparison, has – like a child – attained rights with very few corresponding responsibilities. But unlike a child, those who are in positions of high authority have attained their majority; as one becomes older so too does the burden of responsibility increase to a superior level. Unlike a child, those within the corporation are aware and understand the concept of responsibility. There is no parent upon whom to rely, no other entity who can bear the brunt of the consequences of damaging or destructive practices. It would be logical to conclude that corporate responsibility lies squarely at the feet of those who engage in corporate activities. It is those humans who, like the thief, decide what action is to be taken, yet where the action taken is criminal, all too often those humans are not answerable in a criminal court.

Responsibility is an obligation to behave correctly, a state or fact of being accountable, or having a duty to ensure the well-being of persons and/or other beings. The spectrum of responsibility is dependent on the role assigned to a natural person within an organization, ranging from office junior up to the level of CEO. Ultimate responsibility lies with the individual, not the functional and abstract corporation.

A Lowering of Standards: Corporate Manslaughter Act 2008 (UK)

The crime of corporate manslaughter as a crime was introduced in the UK in 2008. Trumpeted as a piece of radical legislation, it was claimed that the Corporate Manslaughter Act 2008 would make it easier to prove corporate manslaughter allegations and corporations would be held to account for their senior management's failings.

In reality it was the creation of compromise law. The fictional person, unable to be imprisoned for being responsible for a fatality, could instead be fined. It was a prime example of diverting responsibility, in particular superior responsibility, away from the directors and senior managers. The protection afforded was not to the employees and their welfare, but to senior management and their risk of being prosecuted. Instead of a senior manager being prosecuted for failure to uphold his superior responsibilities, the corporation is prosecuted and fined. Loss of human life due to another human's failing is not addressed unless a particular gross breach can be attributed to a specific individual, one who has 'a directing mind.' Thus, by imposing the primary responsibility upon the fictional corporate body, personal liability is evaded. In reality, corporations are not exposed to any greater risks than they faced previously, nor is there a risk of any greater or different punishment than before. As a law to govern corporate activity, the Act has had precisely the opposite outcome. Instead of imposing superior responsibility, it is neatly sidestepped. Thus, the corporate structure and its fictional rights remain in place.

In reality the Act is nothing more than window dressing. It does not create anything new of substance, and it fails to impose the appropriate duty of care on senior officers in the workplace.

Criminal law imposes additional duties which over-ride any obligations a CEO has to the company's shareholders to ensure it makes a good profit. Criminal law imposes a superior governance over humans in all spheres of activity and not just business. For instance, a general must uphold certain responsibilities over and above all military activities. Whilst a general has the right to kill, it is a right that is restricted to certain circumstances. Thus, superior duties and obligations can restrict the application of a right, in some cases prohibiting the exercise of a given right. The general's rights are restricted by law – crimes against peace and laws of war, laws that govern what can and cannot be done within the context of war, and accordingly they set out the boundaries of acceptable conduct. Similarly, with reference to climate change, it could be said that a superior responsibility is owed by leaders of less affected nations to assist nations most at risk of severe climatic disruptions. Written laws are not the only determinant of what those superior responsibilities are; they can be inferred by way of moral obligation, as was demonstrated by the Nuremberg trials. Thus, the absence of a written law at the time of the commission of the crime did not prevent findings of breaches of responsibility subsequently.

Two fundamental points: crime is committed by humans and crime is either a breach of the law and/or a breach of a moral precept. In criminal law the application of rights and responsibilities applies to humans, not fictional persons. It is a recognition that only a human can evade a responsibility, duty or obligation and correspondingly it is the individual who is held to account in a criminal court of law. Responsibilities can supersede rights, especially where those rights have been imposed to gain leverage for a fictional entity. Where rights and responsibilities come into conflict, it is the moral conduct that comes first. In the event that the conduct is found to fall short of what is deemed to be acceptable, it is the breach of a person's responsibilities over the assertion of rights that determines a finding of guilt. Thus, just as a general cannot argue his right to engage in military conflict as and when he so desires, neither should a CEO be able to hide behind the assertion that it is his right to pursue profit freely in whatever way he so desires. A CEO has the right to profit, but this cannot be without appropriate imposition of over-riding duties, obligations and responsibilities.

Superior Responsibility

In international law there is a principle called 'superior responsibility'. It applies in international criminal cases to military commanders and others in high positions who have superior responsibility for decisions taken, either of their own making or by their subordinates. They are in a position of superiority and that carries with it superior responsibility. Where a superior knows or has reason to know that subordinates were about to commit or had committed a crime, he/she is under an obligation to take preventative measures. Those with superior responsibility are those who, in a spectrum of culpability, are at the command and control end of decision-making. The principle is founded on the recognition that human duties and obligations increase proportionately with rank and are therefore commensurate with the greater remit of one's role, whether that be in commerce, politics or the armed services. It is a counterbalance to the additional rights that are granted to those in command. It is a recognition that those beneath and within the chain of command may be acting on orders, which whilst this does not excuse their culpability, does mean that those who are in the driving seat are deemed to have weightier responsibility.

The courts established at the end of World War II in Nuremberg lay the foundations for all international criminal courts. They were not only put in place to hold high-ranking military officers criminally liable under the doctrine of superior responsibility. It was a specific targeting of the acts of individuals and as such the principle was applied also to civilians including industrialists and senior government members. Persons in positions of authority, the courts determined, by dint of their position or rank, carry a higher responsibility that is superior to those for whom they are in charge. The Nuremberg trials also established that persons in secondary positions could not use the defence that they were only following the orders of their superiors. Where the defendants 'should have known' of such violations yet failed to intervene, they were found to be criminally liable. Senior ranking army generals as well as senior members of government and industry all have superior responsibilities.

Outside of international courts and periods of conflict, superior responsibility as an operative legal concept does not simply disappear. An army general will continue to have vested in him superior responsibility for the duration of his military service, be it during times of war or peace. It remains with him for the duration of his role and rank as general. The hierarchy and the execution of decisions of a government or corporation are not in operational terms dissimilar to that of an army and therefore it makes sense that the same concept of superior responsibility applies to a senior government official and senior members of a corporation. Chief executives, directors and senior managers all carry superior responsibility for the duration of their tenureship within a corporation.

A head of a corporation is of course not engaged in military conflict, but is engaged in profit making. Nevertheless, that is not to say there is an absence of knowledge that the company's activities will cause damage, it is merely that it can be a secondary consequence arising out of the primary pursuit. In other words, the destruction is not an end in itself but can be an outcome of the business activity. Knowledge, even without specific intent or direct knowing, is still knowledge that can be imputed. This is the concept known within the law as 'the directing mind', and this means either a person in a position of superior responsibility knows, or should know, of the consequences arising out of any business activity. Superior responsibility implies an even greater duty of care to ensure that any damaging consequences are prevented.

STRICT LIABILITY

The body corporate on the other hand, as opposed to a director who is a human person, is of course a 'fictional person.' As a fictional person the body corporate is not capable of knowledge or intent. A company – an abstract and fictional entity, as opposed to the people within it – cannot have knowledge imposed. Damage which is caused by corporate activity, such as pollution, is thereby governed by what is called strict liability. Knowledge is irrelevant within the legal concept of strict liability: all that is required is evidence that damage has occurred and is causatively linked to the company. Moreover, strict liability environmental pollution crimes are recognized to be so serious in nature that a directing mind is not required to establish guilt. In the words of Lord Bingham, the former Lord Chief Justice of England and Wales:

> Parliament creates an offence of strict liability because it regards the doing or not doing of a particular thing as itself so undesirable as to merit the imposition of a criminal punishment on anyone who does or does not do that thing irrespective of that party's knowledge, state of mind, belief or intention. This involves a departure from the prevailing cannons of the criminal law because of the importance which is attached to achieving the result which Parliament seeks to achieve. [3]

For instance, to prove a water pollution offence in the UK, all that is required is evidence of the damage done, not the intent of the company to pollute. Equally, the offence is deemed so serious that it is absolute: a company pollutes the waters, whether by accident or design, and if caught, will be subject to penalties. There may be additional offences that name some of the company's employees but only where it can be established that he/she had knowledge of the risk, or has responsibility for an employee who should have known of the risk. By way of example, a building site has toxic waste that is not responsibly disposed of and as a consequence leaches into local waterways; the company is immediately at risk of being fined for the pollution created. In addition, where senior members of the company had knowledge of the waste and had failed to remediate the land, those natural persons will also be at risk of prosecution. The difference is the human can be at risk of imprisonment; the fictional person cannot be locked up and will only be at risk of having a fine levied against it.

This is an important distinction: whether a company's directors and senior managers have knowledge or should have knowledge of a damaging activity. The corporate body itself has neither. Company X, as a fictional person, does not think, know or act. Charges brought against Company X are strict liability offences. Thus, when criminal charges are levied against a company (note, not the directors or officers) an anomaly arises: it is the fictional person who is charged, and no-one stands in the dock. Instead, pleas are entered by the lawyer on behalf of the company. Because a fictional person cannot be imprisoned, where there is a finding of guilt, punishment is simply confined to the levying of a fine, compensation and/or an award of costs.

The Impossibility of Holding a Piece of Paper to Account

Following the logic that a corporation cannot have a criminal mind, neither can a corporation have a responsible mind. The corporation as a fictional person does not have a mind whatsoever. After all, unlike a natural person, a fictional person is not capable of thought.

However, as the case of *Santa Clara County v. Southern Pacific Railroad* demonstrated in Chapter 2, the fictional person has in recent times been bestowed with rights, just like a natural person. The consequences of this decision have been enormously destructive. Rights and responsibilities can be withdrawn from a human by placing that person in prison, yet a fictional person cannot be incarcerated. Corporate responsibility exists in limited form, as it does not exist in the fictional person but is in truth vested in the individuals who make up the corporation. The corporation itself is merely a carrier, an artificial body which allows the rights and responsibilities to be passed over in perpetual succession whilst the persons who inhabit the roles changes.[4] Thus, where a corporation has been granted permits to build in a certain place, for instance, in truth it is not the corporation who holds the right to build per se but the individuals who carry the burden of responsibility. The permits themselves are the written documents of evidence that the right to build has been granted to individuals who compose of the company.

The problem is, by applying the incorrect rule, that the corporate body has rights and responsibilities, law has hit a wall. Where those rights and responsibilities are abused, there is little remedy, save for pecuniary penalties, which do little to stem damaging activities. In hiding behind the corporate fictional person, CEOs and directors of corporations have absolved themselves of their responsibilities. Without the ability to impose truly punitive criminal sanctions upon the fictional person, those responsibilities cannot be effectively governed. To accept the fiction of *Santa Clara* ignores the truth: it is the natural person who holds both the rights and responsibilities, not the fictional body and it is the natural person who is accountable in a criminal court, not the corporate body. It is after all the natural person who has the duty of ensuring steps are put in place to prevent pollution occurring, not the piece of paper that gave existence to the fiction.

The legal fiction of corporation as a fictional person is not wrong; indeed it is very practical. But the elevation of its status to that which is equal with a human is wrong; a corporation is literally created out of paper, without the body and mind of a natural person. All rights vested in a corporation can be shifted to the correct legal person by fully imposing those rights upon those who have taken on the positions of superior responsibility. Where a company is found guilty of a crime, it is for the directors to shoulder that responsibility entirely and be subject to criminal and civil sanctions. In doing so, superior duties and obligations would be taken far more seriously by those who take on the position of presiding over a company. Where there is a personal risk of imprisonment for placing others at risk, the impetus of proceeding along a potentially destructive course of action is greatly tempered. Quite simply, in this manner, the appetite for pursuit of profit would immediately be contained and restricted to activities that do not cause damage or destruction. By realigning corporate responsibility with its proper proponents, and holding to account those who hold superior responsibility, the concept of limited responsibility can finally be put to rest.

A compromise in the context of law is the lessening of standards, a weakening of a principle – the principle here being the burden of responsibility. Charging a company ultimately carries with it a compromise by holding a corporate body to account even though it does not have a physical presence before a court. It is the obvious problem of putting an abstract entity on trial; sentencing is limited to a financial penalty because there is no person to lock up. The 'directing mind' could in theory be tried in addition to the company, but the reality is that the more complex the corporate structure the less likely that is going to be achieved. By way of example, the Lyme Bay kayaking tragedy in 1993: the deaths of four teenagers resulted in the conviction of the owner of the activity centre. The company was so small that the individual who ran it was indistinguishable from the company, which meant it was a relatively easy exercise in linking both the company and the directing mind. The owner of the company should have known of the dangers and taken proper steps to prevent the deaths of the four youths. The owner was imprisoned for two years.

The task of identifying the 'directing mind' is intrinsically harder to establish when the organization in question is a multinational corporate structure that has trans-boundary jurisdiction. In 1989, the Exxon Valdez oil tanker struck Bligh

Reef and spilled 10.8 to 32 million US gallons which covered 2,100 km of coastline and 28,000 km^2 of ocean. It was accepted by the courts that the directors had no knowledge (nor that they should have known) of what was being done in Alaska, including that the radar station at Valdez, Alaska, that was responsible for monitoring the location of tanker traffic in Prince William Sound, had replaced its radar with much less powerful equipment that could not monitor the location of tankers near Bligh Reef. Coast Guard tanker inspections in Valdez were not done, possibly because of staff cuts, and spill response teams and equipment were not readily available, which seriously impaired attempts to contain and recover the spilled oil. Liability was limited to punitive damages only, to be paid by the company.

It is only when those in charge of a company are charged as individuals and found guilty that they are adjudicated to be criminals and are subject to possible imprisonment. But, as case law over the years demonstrates, the larger the company, the less likely it is that charges are put (or if put, succeed) directly against senior officers. Herein lies the compromise: in law no one person, or group of persons, takes overall responsibility for the company. The larger the company the easier it is to evade liability, and therefore responsibility. This is a legal loop-hole that has grown so wide as to be an enormous chasm that business has relied on since the day *Santa Clara County v. Southern Pacific Railroad* determined that the fictional person has rights akin to a natural person. Law has grappled with this flaw by accepting a compromise: whilst rights have been accepted, responsibilities have been severely limited. Laws governing corporate activity throughout the world have been built on this compromise for over a century without remedy. The larger the corporate activity the further the scales of justice are tipped out of kilter. In addition, corporate rights have become disproportionately weighted in favour of the corporate body, with the corresponding responsibilities weighing very little. By way of analogy, it is a case of gold ingots placed in scales, weighed against a feather and accepted as a fair exchange.

CORPORATE SOCIAL RESPONSIBILITY

Contribution of net profits into community projects is one example of business engaging in social responsibility. It's an ancient practice, with roots in the concept of tithing, which is derived from the Hebrew practice of *asair*, meaning 'to give the tenth part of' all monies made. Some cultures still honour the principle of tithing, whether it be in the form of giving a percentage of their food to others or in financing community causes. It is a form of giving back into the community, based on the premise that business cannot flourish without the community it serves. In this way, the community gains the benefit of input from those businesses which are reliant on their support. It's a two-way process where both business and community are valued and as a consequence provide for each other. This is an early example of what we now refer to as Corporate Social Responsibility (CSR) – a system that worked well for a long time without the imposition of law. Because it was standard practice, no laws or check and balance systems were required to ensure its application. Businesses gave 10% of their profits and discharged their obligations to society, without questioning their responsibility to the community. Today, were one to read the Annual Financial Report of any given corporation, rarely does one find that even 1% of net profits has been put aside for investment in their communities' projects. To reinstate this practice would go a long way in re-establishing the concept of true Corporate Social Responsibility. One organization that is making good headway in promoting this practice is *1% for the Planet*, an alliance of businesses that donate at least 1% of their annual revenues to environmental organizations worldwide. [5]

When business fails in its duty of care to society it is the role of government to step in to ensure that corporations are prevented from harming the broader social good. There are two ways of doing this; by imposing regulation or by consenting to corporate self-regulation. One is mandatory, the other voluntary. Laws and regulations provide society with the mechanisms in which enforcement can be imposed upon businesses that fail to conduct themselves responsibly, whereas self-regulation is ultimately unaccountable in a court of law. When self-regulation is voluntary, little recourse to enforcement exists. Like all policy, if not backed up by law, it is merely an unenforceable expression of intent. Law, on the other hand is mandatory and, when affirming or admonishing a specific behaviour, decisive. Where companies have demonstrated that they are unable or unwilling to uphold their responsibilities, voluntary codes and practices are insignificant in altering the status quo.

Recent examples of laws that have effected dramatic change in the workplace include Denmark's amendment of their Danish Financial Statements Act, making it mandatory for Danish state and private companies and investors over a certain size to include information on CSR in their annual financial reports. Nearby, Norway amended their Public Limited Companies Act to affirm gender balance on company's boards. The Norwegian government had agreed with the private business sector not to apply the rules if the desired gender representation was achieved voluntarily. After only 13.1% uptake, the law was extended to the private sector one year later with a short two year transition and dissolution of those that failed to comply.[6] The change it effected was rapid and effective. Meanwhile in the UK, at the 11th hour, the then Labour government vetoed mandatory environmental responsibility provisions in favour of a compromise. Section 417 of the 2006 Companies Act was reduced to merely *'Directors to **take into account** social and environmental impact... and include in annual report'*. No reporting standards or sanctions for compliance failure were implemented, rendering the section toothless. Uptake has been nominal (emphasis added).

The issues surrounding government regulation, say industry, pose several problems. Regulations inevitably cannot cover every aspect in detail of a corporation's operations. To do so, they say, would lead to a burdensome legal process, bringing with it debate over interpretation of the law and its grey areas. Instead, an international standard (ISO 6000) for CSR is in the process of being drafted up by industry. It will be voluntary to use, will not include compliance requirements and will thus not be of a certification standard. Much as some of the content is to be commended, until laws are put in place, governance of corporate responsibility remains marginal and firmly outside the reach of the court door.

The same is true for the money that flows into business activity; only voluntary guidelines exist for Socially Responsible Investments (SRI), such as the Principles for Responsible Investment for investing entities. Whilst fundraising activities, community volunteering and voluntary guideline accreditations improve the public perception of a company, economic and legal mechanisms are vastly more powerful.

Today business is perceived to be at odds with non-economic social values despite interest in business ethics accelerating dramatically during the 1980s and 1990s, both within major corporations and academia. Commitment to promoting non-economic social and environmental values under a variety of headings (e.g. ethics codes, social responsibility charters) is to be witnessed on the pages of most major corporate websites. Application and analysis of Corporate Social Responsibility (CSR), a term coined in the 1970s after many multinational corporations were formed, has mushroomed into a business in its own right. As public awareness grows the spotlight shines ever brighter, exposing the damaging and destructive practices that are no longer tenable.

Denmark: CSR as Law

In 2008 Denmark passed the 'Act amending the Danish Financial Statements Act (Accounting for CSR in large businesses)'[7] making it mandatory for the 1100 largest Danish companies, investors and state owned companies to include information on CSR in their annual financial reports. The reporting requirements became effective on 1st January 2009 with information to include: information on the companies' policies for CSR or socially responsible investments (SRI), information on how such policies are implemented in practice and information on what results have been obtained so far and managements expectations for the future with regard to CSR/SRI. Danish industry is now forced to take an active position on social responsibility and communicate it publicly. The Danish government voted in favour of using law to improve the international competitiveness of Danish trade and industry. It is the first country to use law as a mechanism to enforce ethical responsibilities upon corporations. The oft-stated belief that additional regulation will place a burden on the nation's economy has proven to be unfounded.

Norway: Women in Business

As a party to the European Economic Area Agreement, though not a member of the European Union, Norway is required to adhere to the EEC Directive on the Equal Treatment of men and women. In 2002, in compliance with the Equal Treatment Directive, the Norwegian government took affirmative action and voted for a minimum of 40% women to be sitting on company boards of state owned and inter-municipal companies. The law came into force in January 2004, with a two year transition period. Within that time 39% was achieved. By 2010, 44% of board-member positions in Norway were occupied by women (in the UK, whilst there is a concerted effort to highlight the issue, quotas do not yet exist), which suggests that many organizations simply have not addressed gender diversity in the workplace.

The corporate sector publicly criticised the decision, fearing there would be no women capable of taking on the responsibilities of board directors. Some viewed it as a disagreeable and unwelcome intervention by the state in the private business sector.

In response to these concerns the Business Federation of Norway was tasked with setting up a recruitment program called Female Future. Databases were compiled and the National Public Investment Fund provided training courses around the country. Companies were therefore able to find and recruit capable women.

Now very few question the decision to have gender diversity of those in a position of superior responsibility. Norwegian corporate leaders say that this was a necessary reform, that the law opened up their eyes to the talent they were missing out on, and that in the end the women were not hard to find. Businesses just had to look outside the 'old boys' club' and traditionally male boardroom networks. Gender supporting law and policies (such as encouraging flexible working, etc.) at senior level not only create a much more appealing place to work; they have proven to demonstrate sound economic sense as well. Gender quotas for boards have been imposed in Belgium, Iceland, Italy, the Netherlands and Spain.

There is a correlation between imbalance of masculine and feminine in the boardroom and our national and global economies. Barriers to female advancement to board level, as listed by Catalyst, include women's lack of management experience, women's exclusion from informal networks, stereotypes about women's abilities, a lack of role models, a failure of male leadership, family responsibilities and perceived naivety when it comes to company politics. In Norway, women now have an incentive that is embedded in law - even when other priorities start to make demands on their time and energy. In contrast, of the 500 largest corporations in the US, only 23 have a female chief executive officer, or just 4.6%. In the not-for-profit sector women make up 2/3rds of workforce but in the US only 19 per cent of charities have a woman CEO. Whilst 25% of FTSE 100 companies now have women directors, they are primarily in non-executive positions, with many of these serving on several boards and the proportion of female executives remains pitifully low at 9.5% in the FTSE 100, 5% in the FTSE 250.[8]

Norway's boardroom law was radical. Parliament's intention was to make sure that boardrooms should cease to be a male reserve. It was a decision to eradicate sex discrimination at the very highest level. In 2003 the boards of Norwegian Public Limited Companies were only 7% female. By implementing a law with both supporting mechanisms and suitable sanctions, namely dissolution of a company found not to be complying with the law, compliance was a success.

The general public, who are the major stakeholder, are generally supportive of laws that benefit society. Norway's affirmative stance on gender diversity, in the eyes of the public, was within a very short period of time accepted as the norm. Likewise, the implementation of flexitime laws in the UK in 1998 brought a rapid sea-change of attitude in the workplace. Prohibition of smoking, implementation of affirmative laws such as flexitime or gender diversity in the boardroom are all examples of the speed and ability of law to change the course of corporate behaviour.

Norwegian businesses now provide mentoring schemes which increase the visibility of role models and provide access to others for assistance. Mandatory gender equality at the top end of decision-making has yet to make it onto the statute books in the UK. In business, 24 of the FTSE 100's executive directors are women. But this begs the question - which companies have few or no women at the top? Most of them are the Carbon Majors [9]. Women's under-representation in decision-making positions remains prevalent throughout the EU (and elsewhere too), both in politics and the economy, which suggests that policy backed with law could, as seen elsewhere, remedy the imbalance swiftly. In 2011, the European Commission launched the 'Women on Board Pledge for Europe'; in 2012, a proposal was adopted for a quota-based directive (legislation that sets out a goal that all EU countries must achieve. How that is done is for each country to decide) for gender balance on corporate boards. Its main features included a minimum objective of 40% of non-executive directors to be reached by 2020 for companies listed on stock exchanges and 2018 for listed public undertakings.

Now the European directive has removed it's mandatory quotas, preferring voluntary selection mechanisms. Despite evidence from Norway, the directive will no longer set out targets. Instead it obliges member states to establish a process that improves the gender balance in company boardrooms within a limited time-frame.

Mandatory quotas, the evidence shows, accelerate deployment. Norway demonstrated just how fast and effectively law can rectify existing imbalances. For the biggest companies of all, the Carbon Majors, there are hardly any women at all at board level. Without mandatory laws, companies determined to continue with business as usual can.

Chapter 8

ENVIRONMENTAL SUSTAINABILITY

W HISTLE BLOWING, the act of an individual or organization who speaks out about those who are engaged in illicit or damaging behaviour, can often be a first notification that certain activities are causing harm. It is a necessary system of holding governing bodies to account. To ignore the minority who blow the whistle is to become complicit with those who are engaged in destructive processes. Turning a blind eye is a passive acceptance of injustice, which if not halted can become endemic. Effective whistle blowing procedures to halt damaging acts eliminate opportunities for corruption to take root. It also places upon governing bodies a duty of care to ensure all concerns raised are properly examined and, where such activities are established to be in conflict with the primary obligation to foster environmental sustainability, remedied. Implementation of effective whistle blowing procedures, which can be used by those inside governing bodies as well as the independent citizen, imposes upon any governing body a duty to prevent any entity causing damage from the outset. Where systems are not in place internally to protect individuals who speak out, the other route is to go to court.

Rules of Procedure for Environmental Cases

For the first time, a nation has decided to use laws to give protection to eco-whistle blowers. On the 29th April 2010 the Republic of the

Philippines Supreme Court took a radical step. Landmark Rules of Procedure for Environmental Cases were unveiled which include the protection of individuals who seek to bring a case in court to force government agencies to act on their duty to protect and preserve the environment. Establishing the Rules protects persons wanting to enforce environmental rights. Citizens can now file a case, known as a *'citizen suit'*, and defer payment of filing fees until after judgment. Under the new Rules, the *'precautionary principle'* imposes actionable legal obligations and duties on companies engaged in environmentally harmful activities. Citizens can now seek various remedies including an Environmental Protection Order (EPO) for territories at risk or being subjected to harm. Restrictions that normally prevent a citizen from proceeding have been removed.

The Rules are the outcome of the Supreme Court Forum on Environmental Justice which was held in April 16-17th 2009 with simultaneous video-conferencing and open public participation at three universities. It was hailed as an enormous success. The Forum enabled the Judiciary to receive inputs directly from the different stakeholders to the justice system, primarily aimed at determining ways on how the courts can help in the protection and preservation of the ecology and habitat of their territories.

The Philippine Constitution already has in place duties of the state to uphold the citizens' rights to well being, to health and to a balanced and healthful ecology.

Section 15, Article II:
The State shall protect and promote the right to health of the people and instill health consciousness among them.

Section 16, Article II:
The State shall protect and advance the right of the people to a balanced and healthful ecology in accord with the rhythm and harmony of nature.

The duty of the state to protect the people's right to health and a healthy ecology, whilst having been put in place as a law, had until now largely been unenforceable in court by the public or by NGOs and charities with limited funds. High costs and the threat of being at the receiving end of vexatious litigation mounted by corporations was silencing the eco-whistle blowers and citizens who would otherwise speak out. A citizen can now apply for a writ of continuing *mandamus* (compelling another to do something) and other compatible writs, including the issuance of a Temporary Environmental Protection Order (TEPO) as an interim emergency remedy prior to the issuance of a writ itself.

The courts now have the power to protect citizens against Strategic Lawsuits Against Public Participation (SLAPP). Under the rules, persons or organizations can immediately file a defence against a SLAPP claiming harassment or oppression for eco-whistle blowing.

> Section 3:
> *The court shall grant the motion if the accused (the citizen) establishes in the summary hearing that the criminal case has been filed (by the filer) with intent to harass, vex, exert undue pressure or stifle any legal recourse that any person, institution or the government has taken or may take in the enforcement of environmental laws, protection of the environment or assertion of environmental rights.*

Seeking environmental remedy through the courts rarely happens in the UK. Very few environmental cases are fought by citizens and NGOs: this is because a large percentage of potential cases fall at the first hurdle when faced with the risk of a huge bill of costs.[1] In a recent judgment[2], The Aarhus Convention Compliance Committee held that the United Kingdom has failed to uphold its obligations, including its failure to implement a transparent and consistent legal framework to ensure that court costs are not prohibitively expensive. They recommended that the UK government amend their Civil Procedure Rules and that current cost rules be changed, which as they currently stand can force claimants to cover their opponents' prohibitive legal fees, as well as their own and the court's costs. The recommendation now opens the door to the UK implementing similar Rules of Procedure for Environmental Cases. Empowering the individual to act, seeking judicial redress which is prompt and effective, causes a shift in the interests of business away from destructive practices. By placing companies and government agencies in the position of having to justify their activities in a court of law, which may render them subject to unfavourable judgments, imposes an obligation on them to mitigate against any violations occurring in the first place. Implementation of the precautionary principle shifts the burden of proof away from the citizen to the corporation to disprove the assertion that their activities are damaging to the environment and humanity.

FOUR KEY RULES ARE:
1. Strategic Lawsuits Against Public Participation (SLAPPs)
The legal system has widely viewed SLAPPs as an example of the use of law for the purpose of intimidation and as a threat to citizen involvement and public participation, hence the name. SLAPPs are viewed as an attempt to privatize and silence public debate. They are retaliatory lawsuits intended to silence, intimidate, or punish those who have used public forums to speak, petition, or otherwise move for government action on an issue. The issuing of SLAPP proceedings against someone who has spoken out has what is known as a 'chilling effect' on public speech and involvement by frightening others into cooling their support and freezing out any further whistle blowers.

SLAPPs can apply to a variety of different types of lawsuits, including those claiming libel, defamation, business interference and conspiracy as well as criminal prosecutions. Professors George W. Pring and Penelope Canan of the University of Denver coined the term in 1984. They defined SLAPPs using four criteria:

[SLAPPs]
1. involve communications made to influence a
   government action or outcome,

2. which result in civil lawsuits (complaints, counterclaims, or cross-claims),
3. filed against non-governmental individuals or groups,
4. on a substantive issue of some public interest or social significance.

American Judges now dismiss the majority of SLAPPs as a violation of constitutional rights, under the First Amendment which establishes *the right of the people ... to petition the Government for a redress of grievances.* However, in those cases where a SLAPP is not quickly dismissed, the expense of the litigation for SLAPP defendants, both in time and money, often serves as punishment itself and dissuades individuals from speaking out in the future. Individuals who have been hit with a SLAPP — or 'SLAPPed' — often report a feeling of having been sued into silence and feel dissuaded from participating in public life again — quite often the very effect intended by the SLAPP filer. Although a SLAPP filer usually loses in court, he or she may achieve the goal of silencing future political opposition.

The Philippines rules of procedure mitigate SLAPPs in quite a different manner. Instead of the defendant having to fight for their constitutional right to be upheld, the SLAPP filer has duties and obligations imposed which have to be shown demonstrably to have been adhered to. Although the anti-SLAPP provision states it specifically applies to criminal cases, it is anticipated it will be applied in civil cases also. The filer will be far more reluctant to SLAPP a campaigner when the very activities being objected to can, as a result of their own court action, become the subject of an Environmental Protection Order. Because of the defendant's newly acquired ability to lodge a counter-claim seeking an Environmental Protection Order and have their costs deferred until determination of the case (which if, an EPO is put in place, will often result in the SLAPP filer carrying the defendant's costs), the one being 'chilled' switches to the filer. Where there is a genuine cause for concern of harm being done, the activist will find the court, in applying the precautionary principle, far more likely to find judgment in their favour. In the UK, individuals, campaigners and filmmakers who speak out against environmental harm and unorthodox animal practices are increasingly being silenced by the threat of or issuance of litigation from those they are taking to task. Without similar provisions to the Philippines Rules, these voices are being denied the protection of the law.

In many respects the Philippines Rules are both radical and instructive. To ensure that the citizen is given the opportunity to bring a case when an environmental injustice is about to or has occurred, the burden of proof has been shifted to favour the claimant. Whilst a claimant still has to present evidence in support of their claim, whether it be in the form of video evidence, written reports from NGOs and/or verbal evidence from witnesses, in the event that the claimant's evidence is not conclusive of harm taking place, the precautionary principle will now be applied. This is of vital importance.

2. The Precautionary Principle

> SECTION 1. Applicability:
> When there is a lack of full scientific certainty in establishing a causal link between human activity and environmental effect, the court shall apply the precautionary principle in resolving the case before it.
>
> The constitutional right of the people to a balanced and healthful ecology shall be given the benefit of the doubt.

The precautionary principle governs two significant changes of approach. Firstly, the onus of proof shifts to the respondent company/agency to establish that the proposed activity will not (or is very unlikely to) result in significant harm. It places a duty on a decision-maker to anticipate harm before it occurs and make sure all steps have been taken to prevent any significant harm occurring. Failure to do so, or inability to prevent harm can lead to an order prohibiting the activity.

Secondly, an additional obligation is imposed on the respondent in the event where the level of harm may be high. The respondent is duty bound to take action to prevent or minimize such harm even when the absence of scientific certainty makes it difficult to predict the likelihood of harm occurring, or the level of harm should it occur. The extent of control measures increases proportionately with both the level of possible harm and the degree of uncertainty.

3. The Writ of Kalikasan (Writ of Nature)

Where environmental damage is of such magnitude as to prejudice the life, health or property of inhabitants in two or more cities or provinces, seeking the remedy of a Writ of Kalikasan can halt damaging or destructive activities with immediate effect until final determination is made. Like an EPO, the summary process leading to the issuance of the Writ of Kalikasan dispenses with extensive litigation and ensures prompt disposition of matters before the court. The Writ of Kalikasan, which is exempt from the payment of fees, is the first legal tool in the world that empowers the man on the street to seek concrete actions for ecological protection from their government officials. Its procedural use is the same as the Writ of Amaparo, but instead of protecting the constitutional right to life, liberty, and security which has been violated or is threatened by an unlawful act or killing, the Writ of Kalikasan protects the constitutional right to a healthy environment. A citizen can now go to court to seek restitution for violations of the Constitution under section 16, Article II when the state has failed to protect and advance the right of the people to a balanced and healthful ecology. The first case has already gone to court. [3]

4. Environmental Protection Orders

An Environmental Protection Order (EPO), directs or enjoins any person, corporation or government agency to perform or desist from performing an act in order to protect, preserve or rehabilitate the environment.

The Philippine citizen can now summarily apply for an Environmental Protection Order either as a free-standing application, in conjunction with other applications or as a counter-claim to a SLAPP, to prohibit an activity from continuing that is causing or is likely to cause harm. In so doing, the individual has in effect taken on the role of steward on behalf of the territory in question. This has a tangential impact: citizen stewardship eases the financial burden of underfunded government environmental agencies who are unable or unwilling to take action of their own volition.

DEFINING ENVIRONMENTAL SUSTAINABILITY

The importance of environmental sustainability is of such significance that it was put in place as one of the eight global Millennium Development Goals (MDGs), one of the overriding objectives identified as top priority for the next thousand years. The 2005 World Summit announced the eight international development objectives that all United Nations member states were to achieve by the year 2015. They include the reduction of extreme poverty and child mortality rates, the achievement of universal primary education, the promotion of gender equality and empowerment of women, improvement of maternal health, fighting disease epidemics such as AIDS, and implementation of a global partnership for development. Number seven set out the objective *'to ensure environmental sustainability'* which was defined as being the 'reconciliation of environmental, social and economic demands', the 'three pillars' of sustainability. The three pillars were construed to be overlapping, not mutually exclusive, and mutually reinforcing.

This definition has unfortunately served to hinder the application of environmental sustainability in global and local governance, as opposed to advancing it. It is a definition which is flawed due to a) the misrepresentation of the true meaning of the term 'environmental sustainability', b) the incorrect identification of the relationship between the environment, society and the economy and c) the omission of identifying our responsibilities.

Ensuring Environmental Sustainability

The Millennium Development Goal, *'to ensure environmental sustainability'* is an aspiration that is not being met. But maybe it's about going beyond sustainability, not simply aiming for the lowest common denominator. To sustain denotes survival at its basest level; move up a notch and we begin to embrace a more supportive approach - a duty of care for the sanctity of life. Erode the base and everything else falters; species die, habitats become contaminated and people lose their livelihoods. Support the base, foster its growth and life begins again to thrive. It is the actions of society and economic investment that are the pillars of support which will – given the right signals – stimulate resilience and ecological well-being.

The meaning of sustainability has its roots in the late 13th century French verb *sustenir*, to 'hold up, support, endure.' Sustainable evolved in the 1610s to mean 'bearable'; by 1965 the term sustainable growth was recorded as meaning 'capable of being endured or maintained at a certain level.' By the 1970s sustainability became synonymous with the idea of supporting a life-giving system. In ecology the word is now used to describe the conservation of ecological balance by avoiding the depletion of natural resources. When utilised in conjunction with the environment, environmental sustainability denotes the long-term maintenance of well being, based on our relationship of providing support for the growth of the natural world and our responsible use of natural resources.

Environment, Society and the Economy

Without the first system functioning the latter two cannot be satisfied. Take the environment out of the equation and society and our economy collapse. The well being of human life is ultimately dependent upon the successful operation of ecological ecosystems, without which our lives are enormously compromised and can in turn lead to death. Loss of environmental ecosystems is the breakdown of growth of the very territory we inhabit and without a functioning natural world, the rest simply ceases to exist. Life itself is dependent upon the harmonious operation of ecological ecosystems, not the other way round.

The problem with using the word 'reconciliation' is that it assumes a relationship based on the tying together of differing constructs. The definition proposed was based on a tripartite relationship, but this has proven to be ill-judged and unsuccessful. Reconciliation of social and economic with ecological concerns results in failure. Something has to give, and usually it is the environment. Reconfigure the relationship – the environment placed in primary position with society and the economy as secondary support mechanisms – and the premise for a genuinely resilient relationship is set.

Refrains from History

William Wilberforce was the British politician who in the late 18th and early 19th century devoted his life to pursuing the abolition of slavery. Wilberforce rightly identified early on that it was pointless to ask the many millions of slave owners to use or abuse their slaves a little less. That would have been a compromise. To do so equates to modern-day encouragements to be more energy efficient. Slaves were, after all, a form of energy. Jevons' paradox, (improvement in energy efficiency causes exponential use of energy) tells us such an approach has precisely the reverse effect on curtailing consumption. Wilberforce's mission was to abolish the use of humans as slaves and have slaves recognized as having the same rights as all other humans. To do that he realized early on he had to go upstream, to the source of the supply, and outlaw the trade. It took his lifetime to achieve his goal. Two days before he died, slavery was finally outlawed in the UK. Within a year, British slave trade companies were profitably trading in other commodities such as tea and china. Despite all the fears that business and the economy would collapse, neither happened.

Three hundred companies at the time were engaged in facilitating slavery. They all fought against the abolition arguing that it would lead to loss of jobs, that it would be uneconomic, that the public demanded it, and that it was a necessity. What they offered was a voluntary compliance scheme whereby numbers could be capped. Hay would be supplied for bedding. Trading of limited numbers, through an auctioning process between the companies, was proposed. Better to leave it to voluntary market forces, they argued and self-governance. It was conceded by a few that conditions could be improved but that the imposition of any laws would be unduly onerous on business. As a final concession, businesses proposed a levying of fines if they were caught exceeding import limits. [4]

Two hundred years later, three thousand corporations have been identified as being the major players in the destruction of another commodity: this time it is the planet. [5] Prohibition of any destructive practices is met with all too familiar arguments: loss of jobs, disruptive economic considerations, public demand and necessity are repeated frequently and loudly. A well-worn refrain is the complaint that the burden of new business regulations would be disproportionate. Instead, we have cap and trade, determination by market forces, voluntary targets and the occasional imposition of a fine if a business is caught exceeding limits.

PREVENT THE WRONG-DOING UPSTREAM

Modern-day mechanisms have failed to solve the problem: on the contrary we now know they have served to escalate matters. Two hundred years ago, all suggestions to curb the burgeoning slave trade were rejected. Today, those very same mechanisms have been implemented to curb use of energy from fossil fuels. Wilberforce fought hard to demonstrate that slavery of humans was wrong, and no amount of limiting of trade was going to remedy the inherent failure to address the root cause of the problem. Pressure from companies to do otherwise was ultimately rejected. In abolishing slavery, parliament refused to compromise. By comparison, today we continue to address the ecocide downstream. Energy efficiency reductions, limits on greenhouse gas emissions, and carbon permits do not stop the pollution at source.

Corporations and the economy, when faced with the risk of

collapse, can reinvent their wheels overnight. Multi-national energy companies have extensive infrastructure, knowledge, experience and the wherewithal to create and implement an expeditious clean energy strategy when pushed by law; polluting energy industries can, when given the right assistance, successfully turn to trade in other goods and services. Corporations comply quickly to new legislative restrictions – calling for new subsidies to ensure successful reinvention and compliance. Wilberforce argued that to ensure success and prevent a downward swing in the economy, three steps had to be taken. One, withdraw the existing subsidies (sugar plantations were reliant on the use of slaves and were heavily subsidized); two, create laws which prohibit the offending trade; three, make available generous subsidies to facilitate trading of other wares. He did not want to see companies go out of business; on the contrary he wanted companies to provide new innovation to replace their earlier commodity. Businesses at the time argued that to prevent them from trading in slaves would be uneconomic and would result in loss of jobs. 'The public demand that they are a necessity' they claimed. These are the very same refrains that are to be heard today from polluting businesses lobbying governments to prevent laws that will thwart their current business dealings. History demonstrates clearly that voluntary mechanisms produce little more than business as usual scenarios. Prevention on the enormous scale that is required is not afforded by merely damaging a little less. That is akin to asking everyone to 'use your slave a little less.'

BANKING ON IT NOT BEING A CRIME
Very few legally binding rules exist to determine what a bank can or cannot invest in. Short of a project being identified as a criminal activity, there is little to curtail the flow of billions of dollars per annum into damaging or destructive projects. Save for banks who take an ethical stance, such as Triodos, most banks are driven by very different considerations. Governance of financial structures are on the whole unaccountable and hidden. The exchange of enormous sums of money are the back-bone of most damaging

and destructive activities. Halt the flow of money into such projects, change the direction of investment into benign ventures and the transition into cleaner solutions is rapid and effective. This necessitates law.

The World Bank, the international financial institution that provides leveraged loans, already has a structure in place that is premised on preventing adverse environmental impacts and financing projects that are sustainable and environmentally sound. Category A projects are projects that carry a risk of the level of harm being high, wide-ranging in geographical impact, long-term and severe. Yet the Category A projects are not prohibited, they just require a more extensive environmental assessment before being signed off. Although there is a provision for public consultation on Categories A and B projects, it is questionable whether proper consultation is effected. These are as good a place to start; tighten them a notch and raise the bar to what is no longer acceptable as an appropriate investment category. Category A is the category that the Deepwater Horizon oil rig project comes under; if Category A were to be deemed illegal, financial investment of similarly dangerous ventures would stop overnight. Investment would then be freed up to flow into sustainable innovation and environmentally sound projects.

The World Bank Operational Policy (OP) 4.01 was drafted up in 1999. Under its guidelines, the World Bank requires environmental assessment (EA) of projects proposed for Bank financing to help ensure they are environmentally sound and sustainable, and thus improve decision-making (paragraph 1). It is premised on the Bank's preference for 'preventative measures over mitigatory or compensatory measures whenever feasible' (paragraph 2). A proposed project is Category A if it is 'likely to have significant adverse environmental impacts that are sensitive, diverse, or unprecedented', and will normally require an Environmental Impact Assessment (or a comprehensive or regional EA). A proposed project is classified as Category B if its potential adverse environmental impacts are site-specific, if few of them are irreversible, and if mitigatory measures can be designed more readily than for Category A projects. The scope of an EA for a Category B project will be narrower. A proposed project is classified as Category C if it is likely to have minimal or no adverse environmental impacts. Categories A and B are to be be subject to public consultation (paragraph 15).

Make ecocide a crime, and Category A projects shall automatically become illegal. Instead, prioritize Category C and subject all projects to proper public consultation. These are radical proposals. They are also ones that are in alignment with existing voluntary guidelines for the most powerful bank in the world. By identifying which type of investment to favour and implementing law that supports the new investment, only then will the international community actively advance sustainable development. Law can close doors and open others. In setting the bar strictly, by closing off one route, it becomes easy to ascertain what is required in its place so that investment pours through the newly opened doorway. Public consultation, proper engagement with the primary stakeholders of a given project, ensures that the wider concerns of the community are placed firmly at the forefront of all investment decisions. Implement this as law and it will actively facilitate the flow of the finance required for a rapid and effective transition to cleaner solutions.

In its 2009 Environmental Impact Analysis for the Deepwater Horizon well, BP asserted it was unlikely, or virtually impossible, for an accident to occur that would lead to a giant oil spill which would seriously damage beaches, fish and mammals.[6] Going by the World Bank categorization, Deepwater would qualify as a Category A project on the basis that it was 'likely to have significant adverse environmental impacts that are sensitive, diverse, or unprecedented.' BP's take on the risk analysis, however, seems to accord with Category C. Subsequent events tell us otherwise. Make Category A projects illegal and when a taxpayer next challenges Stephen Hester, CEO of RBS, as to why he now refuses to invest our money in Category A projects, he will have the opportunity to put the response to the public "well, it's a crime."

Best Practice Procedures

Effective governance of land and water, be it local, national, regional or international, can only work for the benefit of the wider community by placing environmental sustainability at the core of all decision making. It is through the implementation of policies – policies that sustain, prop up and support the surroundings and conditions on which all life depends – that our goal of ensuring environmental sustainability will be met. Systems and policies that are damaging and destructive are inevitably unsustainable for the environment. What those policies sustain is quite a different framework, one that will facilitate only the advancement of conflict, fear, separation and greed. In any decision, one can examine what is being protected – private or public interests, life threatening or life giving systems – then decide which path to take.

Policies in themselves are not enforceable without law; law creates a binding structure within which society can operate effectively. Where territories are at risk, by implementing laws that impose the overriding objective of a duty of care over and above all other contractual rights, the territory can be protected. This means putting in place systems that are premised on understanding our relationship with the natural world as one of stewardship, and identifying our duties as stewards or trustees of our territories. It also means building accountable, transparent and participatory systems of governance at all levels. It is a shift to a governance that protects society and the wider community interests over and above the private interest.

Accountable Governance

The word governance derives from the Greek verb *kubernáo* meaning 'to steer' and was first recorded to be used by Plato. Governance is the making of decisions that define expectations, grant power and verify performance. Politics provides a means by which the governance process operates. Politicians promote policy; when it is implemented the enforceability and the power behind policy usually comes from legislation. Conceiving of governance in this way, one can apply the concept to states, to corporations, to non-profits, to NGOs, to partnerships and other associations, to project-teams, and to any number of humans engaged in some purposeful activity.

Increased Transparency, Bellandur, India

Bellandur is a relatively well off agricultural village near Bangalore, where access to education has contributed to making the village almost 90% literate. But that is not all. 'The project has also been successful because of the active cooperation from the villagers,' said retired public sector employee Ganga Reddy.

Bellandur's e-governance project commenced in 1998 with a single computer installed in the village to replace the old typewriter. The village office now has three computers, funded by donations from wealthier farmers as well as companies that operate in the area. 'Revenue loopholes have been plugged. All the records are available at the click of a button,' according to Mr Jagannath. Ms Shobha, a programme operator employed by the village administration, explains that the opportunity for officials to hold things up by shuffling paper from file to file or desk to desk has been eliminated. 'People can get their land registered in record time. Earlier, it used to take anywhere between seven and 10 days,' she said.[7]

Good governance is all about systems being open to public scrutiny, whether it be big or small. More than anything else, lack of accountability rapidly leads to lack of confidence by the public. When communications break down, dissatisfaction sets in and society begins to fracture. Relationships can easily crumble and endanger the very fabric of the community. On the other hand, where provisions have been put in place to ensure all practices and decisions are open to scrutiny and publicly monitored, less suspicion and mistrust arises. In giving the public the right to hold to account decision makers, something as simple as setting up a transparent system to register documents can prompt a complete shift of balance of power, from a top down approach to a more democratic, decentralized system in which others are able to participate.

PARTICIPATORY GOVERNANCE

Participatory governance focuses on deepening democratic engagement through the participation of citizens in the processes of governance. Citizen engagement in public decision-making facilitates deeper involvement with political issues. Participatory governance also supplements the roles of citizens as voters or as watchdogs through more direct forms of involvement. In doing so, the decision-making is devolved from central control to local administration.

CIVICUS is an international organization that campaigns for civil society engagement. Its key objectives are: 'to amplify the voices of CSOs (Civil Society Organizations), to disseminate to a broader audience CSO demand,; to open up space for CSOs, at all levels and to keep campaigning on the core global issues that affect us all.' CIVICUS have comprehensive toolkits for community governance that cover issues of legitimacy, transparency and accountability issues. Laws and local procedures from all over the world are used to demonstrate best practice. Other toolkits are available from Transparency International, including 'Transparency Tools to Support Transparency in Local Governance', an invaluable resource for those working at improving transparency, combating city level corruption and inefficiency and promoting effective citizen participation. [8]

The Open Public Meetings Act

As a part of a nationwide effort to make government affairs more accessible and responsive, the Open Public Meetings Act was passed by the US legislature in 1971. Procedurally, the Act determines who may attend meetings, how minutes should be maintained, and the procedure for issuing public notice well in advance of a meeting. It recommends that agencies adopt a schedule of regular meetings through ordinance, resolution, by-laws, or by whatever other rule is required for the conduct of business by that body. It also emphasizes that agencies give notice of each special meeting. Glendale City, Arizona, is an example of one city that actively encourages citizen participation, for them it is an essential component of good governance and democracy.

Executive meetings (i.e. private meetings which are closed to the public) are the exception, and can only be employed in very limited circumstances. Discussion of personnel issues and the obtaining of legal advice are the two main exceptions. Confidentiality laws prevent the minutes of an executive session being open to the public. Nevertheless, any citizen of Glendale City can attend or watch their regular council meetings on the city's 'KGLN 11' cable channel and will hear the time and date announced when an executive session has been scheduled. The executive session is listed, the agenda and a summary of what will be discussed can be viewed on the city's web page. [9] The final agenda for an executive session is posted at least twenty-four hours before the executive session is scheduled to begin. Citizens can obtain copies of the agenda for any executive session, including past executive session agendas, by calling or visiting the city clerk's office on the fourth floor of city hall.

Co-determined Governance

The legal landscape will inevitably change to accommodate the internationalization of business at all levels. Company laws of other nations are relevant to not only the largest of corporations but also a rapidly growing number of small and midsize business enterprises. As cross-border business expands, the schism between private and public business law is pulled ever tighter. Global laws for global business practice can go either way – a global, legal duty of care or, in its absence, continuance and escalation of ecocide.

The European Union drew up a radical Directive addressing corporate governance, supervision of corporate management and employee involvement in company decision-making. Entitled the Fifth Directive on Company Structure, it was strongly debated in the 1970s by the then European Community.

Co-determination is a practice whereby the employees have a role in management of their company. The first co-determination laws began in Germany, hence the term (which is a literal translation of the German word Mitbestimmung). The concept of worker participation in management was initially introduced into the coal and steel industries and soon spread elsewhere. Co-determination rights are different in different legal environments, but the essence of the principle remains the same as set out in the proposed Directive in 1974, which set out a mandate that worker representatives hold seats on the boards of all companies employing over 500 people. By way of example, in Sweden, eighty percent of the workforce is organized through the trade-unions which have the right to elect two representatives to the board in all Swedish companies with more than 25 employees. This principle could be easily extended to include an elected community member of the territory where business activities are taking place. Representations from the community thereby become an integral part of the corporate structure.

One of the main achievements of co-determination is that employees are more involved and have more of a voice in their workplaces, which in turn results in a positive return in higher productivity. Industrial relations suffer lower levels of strike actions, days lost to employee illnesses and better pay and conditions are secured for employees.

On the back of the success of German co-determination the European Community (now the European Union) drafted the 5th Directive on Company Law, proposing a similar two-tier structure of a supervisory board and a management board with worker representation on the supervisory board. Membership of the supervisory board is composed of one-third elected by shareholders, one-third by the employees and one-third independent, co-opted by the other two-thirds. In Britain proposals for co-determination were also drawn up and in 1977 Harold Wilson's Labour government commissioned a research paper, named the Bullock Report. External events intervened to prevent the report's recommendations to roll out co-determination from ever being implemented. The UK government's attempt to control inflation and the beginnings of Thatcherism brought an end to co-determination in the UK workplace. In Europe, the momentum was lost and the 5th Directive on Company Law was never brought into force.

Implementation of co-determination increases the power of the employee, extending their say in the decision-making process. A company therefore becomes by necessity accountable to its immediate community and is in turn held to account by its own employees. What has been discovered is that it is for this very reason that firms rarely adopt co-determination voluntarily. [10]

Include representatives of the wider community and the potential for full and fair representation of all stakeholders can take place. Representation by either of the three ways set out above guarantees that more informed decision making is equally applicable in micro and macro situations. The key is to provide for and empower the largest stakeholder of all – the community – to be equal participants in the decisions that affect the lives of those they represent. In legal terms, company law on this basis gives for the first time proper substance to the term 'the triple bottom line: people, planet and profit'.

EFFECTIVE RESOLUTION:
FIRST RESTORE, THEN ORDER A FINANCIAL PENALTY
Each day a hundred living species become extinct, 1000 acres of peat bogs are excavated, 150,000 acres of tropical rainforest are destroyed. Each day 2 million tonnes of toxic waste are dumped into our rivers and seas, 22 million tons of oil are extracted, 100 million tonnes of greenhouse gases are emitted. [11] Today large-scale habitat destruction, massive soil depletion, extensive deforestation all lead to worldwide disruption of natural cycles and the irreversibility of extinction. Now instances of mass extinction occur with greater frequency, greater rapidity and greater impact than at any other time. This is ecosystem destruction on a phenomenal and unprecedented scale.

When land, natural resources, and labour are regarded as commodities that can be sold in the market place, corporate enterprise separate themselves from the concerns of other humans or their environment. Instead self-interest and money-making becomes the obsession, and as demands grow so too does business, all the time strengthening and reinforcing the belief in the existing structure.

The meaning of resource was first recorded in the 1610s, as a 'means of supplying a want or deficiency.' A natural resource is, of course, far more than that: it is a home for various ecosystems where biodiversity can flourish. When we deplete our stocks to satisfy our wants, we restrict nature's capacity to recover to its original state and replenish the rapidly dwindling assets.

The Kiev Protocol on Pollutant Release and Transfer Registers (PRTRs)

The Protocol on Pollutant Release and Transfer Registers was adopted at an extraordinary meeting of the Parties to the Aarhus Convention on 21st May 2003. It was adopted by thirty-six member states who signed the Protocol in Kiev. It is a legally binding international instrument with the stated objective "to enhance public access to information through the establishment of coherent, nationwide pollutant release and transfer registers (PRTRs)." Article 2, paragraph 7 of the Protocol defines release as:

> *any introduction of pollutants into the environment as a result of any human activity, whether deliberate or accidental, routine or not routine, including spilling, emitting, discharging, injecting, disposing or dumping, or through sewer systems without final waste-water treatment.*

PRTRs are inventories of pollution from industrial sites and other sources. Although regulating information on pollution, rather than pollution directly, the Protocol is expected to exert a significant downward pressure on levels of pollution, as no company will want to be identified (it is presumed) as one of the biggest polluters.

The Protocol became international law binding its Parties on 8th October 2009. Although signing of the Protocol is now closed, the Protocol is open for accession by states and regional economic integration organizations constituted by sovereign UN member states which have transferred their competence over matters governed by this Protocol. All states can participate in the Protocol, including those which have not ratified the Aarhus Convention and those which are not members of the Economic Commission for Europe. It is by design an 'open' global protocol.

The Protocol established a Working Group on PRTRs, which was set out in an earlier resolution. Under its mandate, the Working Group is charged with identifying and carrying out activities to be undertaken pending the entry into force of the Protocol, in particular to prepare for the implementation of the Protocol through the preparation of guidance documents and the sharing of information and experience gained. Parties that are regional economic integration organizations, are to be construed as applying to the region in question unless otherwise indicated. It is charged with reporting periodically to the Meeting of the Parties to the Convention on progress made in respect of the ratification of the Protocol and steps taken towards its implementation, and to prepare for each session of the Meeting of the Parties.

RESTORATIVE JUSTICE

In 1971, an American law lecturer, Christopher Stone, posed a question to his students: what if trees had legal standing? It was an avenue of academic and legal enquiry that culminated in his seminal book *Should Trees Have Standing?* Professor Stone suggests that the idea of holding legal rights involves at least three aspects: 'first that the thing can institute legal actions *at its behest*; second, that in determining the grant of legal relief, the court must take *injury to it* into account; and, third, that relief must run to the *benefit of it*.' [12]

It is an important starting point for any enquiry into legal rights. Taking as an example a forest destroyed by a company which has razed the territory to extract fossil fuel, another way of phrasing this would be to ensure: a) that a human can seek legal remedy on behalf of the forest; b) that when determining remedy the court will give proper consideration to the damage, destruction to or loss of the forest; and c) that the forest benefits from the remedy imposed. Stone hits on a very pertinent point;

without benefit of remedy to the forest (or any being in which rights are vested), nothing is achieved. To fine a company found guilty of harming the forest does nothing in terms of restoring the damaged caused in the first place. Implicit within Stone's analysis is the imposition of a duty of care upon the company; the company has to make good the damage, not merely buy its way out of the problem.

RESTORATION: REMEDIATION AND REGENERATION

Restoration is a form of remedy that starts from the premise of putting back or returning something to its former condition; it is a remedy that directly addresses the losses of the beleaguered party to restore that which has been harmed rather than simply fixating on the punishment of the perpetrator. Shifting the focus to the restoration required will inevitably have a cost attached, but the price is not the final determinant, it is the restoration required that has to be adjudicated first. Determination of cases on the premise that the burden is on the company to implement the necessary restorative steps, at whatever cost that may require, is a proper evaluation of remedy. The finance or wherewithal required to do it by the liable company can then be determined after the primary judgment has been given setting out the restoration that has to be undertaken. Restorative justice is built on an understanding of our relationship with nature and the duty to remedy the harm caused.

REMEDIATION

Restoration by way of remediation of land or waters that have been subjected to the extraction of natural resources, is a remedy that is on the whole ineptly and inadequately deployed. In law, remediation means providing a remedy. Environmental remediation is specifically concerned with the removal of pollution or contaminants from environmental media such as soil, groundwater, sediment, or surface water for the general protection of human health and the environment or from a brownfield site intended for redevelopment. Reversing and stopping destructive processes is the first step towards restoration; the next step is to input beneficial systems for future growth.

There are many types of remediation, some benign, some destructive. Where a remediation process creates a secondary legacy of pollution, such as the use of one toxin to break down another, the problem is too often being simply replaced by another, and true remediation has failed. Bioremediation, however, is one particular type of remediation process that looks to nature for the remedy. Microorganizms, fungi, green plants or their enzymes are used to neutralize specific soil contaminants, such as introducing bacteria to break down chlorinated hydrocarbons. An example of a more general approach is the cleanup of oil spills by the addition of nitrate and/or sulphate fertilizers to facilitate the decomposition of crude oil by indigenous or exogenous bacteria.

This is just a first step however. Whilst bioremediation can be used to facilitate decomposition, its effectiveness is in part dependent on the nature, scale and the location of the territory in question. For instance, polycyclic aromatic hydrocarbons (PAH) and total petroleum hydrocarbons (TPH) are two of the most damaging pollutants pumped into the atmosphere from combustion and are ubiquitous in the urban environment, in refinery sludges, waste oils and fuels, and wood-treating residues. TPH and PAH are not only chemicals of concern at many land based waste sites, the presence of such pollutants also exist in water and sea beds. Sediment contaminated with TPH and PAH can be treated by a bioremediation process called 'ripening' which is a soil conversion process that irreversibly converts waterlogged clayey sediment into aerated soil by the action of desiccation and structure development. Land based soils can be treated either on site or removed for safe treatment elsewhere, and shallow waters, such as harbours, can be dredged and the contaminated sediment removed offsite for remediation.

Putting Toxic Tailings on Trial

In July 2010 Syncrude was convicted on one count under Section 155 of the Alberta Environmental Protection and Enhancement Act for failing to provide appropriate waterfowl deterrents at the pond and one count under the federal Migratory Birds Convention Act for allegedly depositing or permitting the deposit of a substance harmful to migratory birds in waters or an area frequented by birds. Provincial Court Judge Ken Tjosvold found Syncrude guilty of failing to stop the ducks from landing on its seven square mile tailings pond on 28th April 2008.

'The opening remarks solidified our belief that this case is really about the toxic tailings lakes themselves being on trial', commented Sierra Club Prairie Chapter Director Lindsay Telfer, who had initially brought the private prosecution against Syncrude after the state failed to act. Judge Ken Tjosvold ruled that Syncrude was indeed responsible for its tailings pond where the ducks were found, and that it 'did not deploy the [bird] deterrents early enough and quickly enough' around the pond which contained the toxic tailings. Syncrude had failed to prevent the deaths of 1606 birds, by depositing a substance harmful to migratory birds in waters with insufficient cannons in place around the pond. It was found that there had only been eight in place compared with 130 the year before.

Sentencing has been adjourned. The company is likely to face fines later this year of up to $800,000 and a separate fine may be levied for every dead bird. There will be no order for remediation or prohibition.

Unlike ponds, which can be remediated, the situation is very different for the seas. Where the sea has been badly polluted, unlike the land, there is no known option yet discovered to successfully remediate large tracts of deep water. Worldwide, remediation of the seas has been non-existent: instead, in particular where spills occur, attempts are focused on the partial decomposition of surface pollutants (often with toxic chemicals), which then sink and in the process enter into the food chain after being ingested by sea-life. Until a suitable remediation of the seas has been discovered, all activity that places the seas at risk of loss of life on a grand scale threatens our fragile ocean life. To undertake any activity that brings with it a risk of destruction is of concern, but where that risk cannot be remedied it calls for a complete ban on such activities.

Where remediation can be applied, it often requires enforcement, either by a governmental agency or by a court order. Remediation of a site is after all an expense that few wish to pay. Those who fail to take responsibility from the outset are leaving a legacy which can remain unresolved for centuries, affecting untold numbers of species in unforeseen ways.

Regeneration

Remediation alone does not ensure restoration. Remediation is just the beginning; it is the removal of the hazard. Regeneration is an equally important measure; the reinstatement of pre-existing conditions that were in place before being destroyed by the hazardous substance or activity. Regeneration is the provision of that which will return the ecosystems of the affected territory back to a state of optimum well-being – literally re-establishing conditions so that life can return. Thus restoration involves both sides of the equation; identifying at source what it is that has to be stopped and removed, then identifying what has to be started and replenished.

Grass roots can stretch metres underground, if left to grow. It is in the roots of grass that the majority of its carbon is stored. Where height is reduced, by foraging animals or by cutting, the plant reacts immediately by shortening its roots. As the dead matter rots away the previously stored carbon gradually returns to the atmosphere. The roots will gradually extend again, if not cropped, but grass that is kept short will not have the opportunity to build up long carbon-rich roots. Partly as a result of this, the ground will hold less water, and almost all vegetation will eventually die. What is known as a 'brittle' area is a territory that has long periods without rainfall, resulting in desertification; overgrazing can tip the land into irreversible decline. [13] Disrupt the cycle enough and the following downward spiral occurs: biodiversity loss leads to reduction of biomass and loss of photosynthesis, which reduces carbon sequestration thereby reducing soil matter, resulting in a domino effect of loss of soil stability, loss of nutrients and loss of hydrological regulation which amounts to systemic dysfunction, including loss of fertility, floods, drought, mudslides, dust storms, food insecurity, poverty and unchecked population growth. Reverse the process and the natural cycle can be regenerated. Simple interventions to remove damaging systems and replace with regenerative processes result in the flourishing of life. Seed grass and it can provide a root system upon which a multitude of biodiversity can attach and grow, providing the basis for an abundance of life to take root. One of the most important and simple forms of regeneration is to let the grass grow.

Loess Plateau: from brittle desert to green oasis in eight years

The Loess Plateau is one of the most hostile terrains on the planet. It covers an area of some 640,000 km² (roughly the size of France) in the upper and middle reaches of China's Yellow River. In early Chinese civilization the Loess Plateau was highly fertile and easy to farm, but by the late 20th century it had become a dustbowl. Heavy rainstorms would wash through the desert, leaving sand to dry and lift in heavy winds which would eventually come to land on cities many thousands of miles away. Centuries of deforestation and over-grazing have resulted in degenerated ecosystems and poor local economies. The Loess Plateau Watershed Rehabilitation Project was launched in 1994 to prevent further desertification and restore the land to its previous fertile state.

The Loess Plateau was restored to a green oasis within eight years. It was a hugely ambitious project in terms of size and community engagement; destructive practices were stopped and ecologically diverse systems were put in place to restore the natural cycles of life. Farming traditions that were recognized to have negative impact, such as the free roaming of goats, were banned and steps that proved to have positive benefits, including tenureship of the land by the inhabitants, have been so successful that they are now applied elsewhere. The land was zoned: high, difficult to reach areas and ravines were designated to be ecological zones to be returned to nature, lower lying lands were designated for agricultural use. The majority of the land was to be returned to nature. The premise behind this was practical. In returning higher lands to biological diversity, heavy rains were not only accommodated but proved beneficial, storing essential water supplies in vegetation, rivers, land cover and forests which retained moisture thereby supplying the low lying areas throughout the dry seasons. Last year the area suffered one of its worst droughts. Despite this, foodstuffs continued to be so abundant that the people of the Loess Plateau were distributing food to neighbouring provinces. [14]

Large-scale restoration of degraded ecosystems such as wetlands, forests, and deserts are tremendously effective environmental and social remedies. For the first time ever on the Loess Plateau, children are entering into higher education rather than being shackled to a life of subsistence farming. Population rates have dropped dramatically due to women being educated; overall health has improved markedly. Quality of life has risen exponentially for all who inhabit the area and dust storms are a thing of the past. Ecologically, the results are remarkable. The creation of biodiversity and biomass has resulted in raised water tables, forestry, soil stability and the creation of one of the largest and most effective natural carbon sinks in the world. Correspondingly, such intervention has dramatically stymied biodiversity loss, fresh water stress, desertification, loss of soil fertility, poverty, population growth, conflict and climate change.

Compared to some tree planting projects, holistic ecological restoration projects such as the Loess Plateau Watershed Rehabilitation Project have been found to be many times more beneficial to immediate and future communities with the added bonus of being capable of sequestering far higher amounts of carbon than a standard mono-crop plantation scheme. It is a remarkable testimony of what can be done to restore the fertility of a large-scale barren land within a very short time-scale, facilitating increased well-being for all who live and are dependent on the area.

Transitioning to Cleaner Solutions

Legislation has huge power to generate transformation overnight. Within just a year of rendering slavery unlawful, traders were profitably trading in other commodities, such as tea and china. All slave trade subsidies had been withdrawn and replaced with subsidies for new trade. More recently, swift transformation was effected during World War II in America when the automobile industry was ordered to apply their skills to the making of planes for the war effort. Between the end of 1941 and early 1945 the US government, business, labour and the public mobilized. By 1943 *'the United States... had completed its administrative apparatus for managing economic mobilization, revised its strategic plan and estimates of force requirements, stabilized its manpower and labor problems, and erected the factories and recruited the workers necessary to pour out the greatest arsenal of weaponry the world had ever seen.'*[15] In May 1940 President Roosevelt arranged for 50,000 new aircraft to be built.[16] Emergency legislation was passed overnight rendering the building of cars illegal; generous subsidies were given and a short transition period of just a matter of months was granted. Engineering training was essential to ensure the planes were properly built and thousands were recruited and trained; the training was reduced from five years down to just seven weeks. Thousands of US planes flew into the air by the end of 1941: by the end of the war the number of planes built was ten times the initial specification.

The organized labour movement, lifting the US out of its Great Depression, became a major counterbalance to both the government and private industry. American industry was revitalized by the war, and many sectors such as aerospace and electronics were by 1945 completely oriented to defence production. The war's rapid scientific and technological changes continued and intensified trends that had begun during the Great Depression, creating a permanent expectation of continued innovation on the part of many scientists, engineers, government officials and citizens. The United States continued to enjoy unprecedented economic and political power after 1945, in contrast to the severely damaged economies elsewhere in the world. Quality of life substantially increased during the war for Americans, leading many to build permanent improvements to their material circumstances whilst fears of returning to a postwar depression rapidly receded.

Mobilization to a new era of stability happened fast for America. We can do the same again today, in every country in the world. However, business cannot do it without the assistance of governments passing laws. Think back to the UK Canal Acts in the mid-18th century. Hundreds of Acts were passed in short succession, freeing up a whole swathe of industrial activity. America mobilized a whole nation in a completely different direction within a very short timescale. Every nation has the power to pass laws overnight when they deem it a state of emergency. New laws can be implemented at international level equally fast. It requires political will to do so for the outcome to be a rapid shift to transition production – instead of defence production – at a scale comparable to war time. The emergency of our time is the destruction of the planet.

The planet has the remarkable ability to regenerate – if we help it can be done with incredible speed and we all stand to benefit. We can study history, taking our cues from past experiences and building on them. The collapse of the climate negotiations in Copenhagen may just have been the best outcome of all. There is now a short moment of opportunity to rethink the solutions, work out the next steps and put them in place swiftly and decisively.

Chapter 9

NEW DEVELOPMENTS

I N 1948 the Universal Declaration of Human Rights was signed on the back of the humanitarian injustices of World War II. It is the document that has led to the recognition of human rights throughout the world. Although not a treaty, the Declaration was explicitly adopted for the purpose of defining the meaning of the words 'fundamental freedoms' and 'rights' appearing in the founding Charter of the United Nations. The Declaration of Human Rights is a fundamental constitutive document of the United Nations. It is a tool which lawyers throughout the world use when seeking justice for those whose human rights are breached. Without it we would not have the capability to seek redress for the human injustices that continue to afflict the world. But now we have to go further. Now we have ecological injustices which are so enormous that they can no longer be ignored. We will require nothing less than an equally embracing declaration to protect the rights of nature.

Non-human Rights

To date there is only one country, Ecuador, that has so far written into its statute books the explicit contention that nature has rights, rights which can be defended in a court of law. To ensure ecological justice is to ensure that ecosystems remain intact and functioning. When we have ecosystem breakdown, resource depletion often leads to conflict and ultimately war. Rights for the wider community are another way of providing protection to our habitat.

The application of rights for non-human beings provides a valid legal construct where violations to the planet are without voice. In the absence of non-human rights, attempts to bring actions to remedy environmental damage and destruction based solely on adverse human impact rarely succeed. Climate change and pollution cases are currently failing at the court door for being argued on the basis of human rights breaches, because of difficulty of establishing a causal link. For instance in a pollution case, where the people allege that the pollution is causing their illnesses and a consequent breach of their human rights, powerful counter-arguments point to the fact that cancer can be caused by many factors, not just by nearby toxic dumping. By imposing rights for nature, the communities are now legally empowered to protect their natural environment and rainforests. A breach of a non-human right, such as the right not to be polluted, would establish a direct causal link. In this situation, a soil or water sample could easily prove the case. Where the crime of ecocide is alleged, the implied right is the right of the wider community, humans and the territory affected, not to be harmed. Thus the causal link to be demonstrated in court can include soil and water damage as well as animal and human harm.

Until relatively recently the prevailing belief was that we, as humans, have superior rights to the exclusion of rights being vested in other sentient beings. Superior rights implies the existence of superior responsibilities. It is the reliance by some on their perceived superior rights without thought for the consequences and without application of their superior responsibilities that has led to our current crisis. We have accepted as the norm laws that have implied the right to destroy and the right to pollute so extensively. Just because it is the norm does not mean it is right. It is the (silent) right which fictional persons have obtained that have created so much damage and, as a consequence, an imbalance in our ecosphere so great that it is threatening to destabilize all of earth and mankind. Law shapes our societies, our way of thinking, our behaviour. By imposing upon land the concept of it being a 'commodity' that can be owned and dealt with in a similar fashion to, say, a table, legal systems legitimize and encourage the abuse of the earth by humans.

The ascribing of rights to non-human species is on the increase. A Universal Declaration of Animal Welfare, promoted by The World Society for the Protection of Animals,[1] has generated global support with numerous countries indicating their willingness to amend their laws in line with the Declaration. As a result of this initiative, animal rights are growing in recognition throughout the world; for instance in 2009 Spain legally recognized rights of chimpanzees. In Europe there are regulations setting out duty of care provisions for animals which the RSPCA rely on to bring private prosecutions. Rights of animals are not always directly specified in legislation such as this, but are implied. In the international arena the United Nations Convention on the Law of the Sea implies rights for the seas. Water rights and rights of the atmosphere not to be polluted are now being explicitly formulated by lawyers and academics.

BILL OF NATURE'S RIGHTS

On Sunday 28th September 2008 the people of Ecuador voted for a new Constitution. Included are legally enforceable rights of Nature. Ecuador has codified a new system of environmental protection based on non-human rights. In a country rich with ecological treasures, including the Galapagos Islands and part of the Amazon rainforest, the constitution also places a duty on the government to implement measures to prevent destruction of ecosystems and species extinction.

Article 71 of the Rights for Nature (Chapter 7) of the Ecuador Constitution reads:

Nature, or Pacha Mama, where life is reproduced and occurs, has the right to integral respect for its existence and for the maintenance and regeneration of its life cycles, structure, functions and evolutionary processes. All persons, communities, peoples and nations can call upon public authorities to enforce the rights of nature. ...

Ecuador's Constitution protects the rights of ecosystems to exist and flourish, the citizen's right to speak on behalf of nature and the responsibility of the government to remedy any violation of an ecosystem right. Ecuador has enshrined in their law well-recognized pre-existing rights, the difference being that now they have legal validity. Only with legal validity can court action can be used to enforce steps to be taken when breaches occur.

Crimes Against Present and Future Generations

A crime against future generations can be intergenerational in outcome – an act (or omission) can take place today that has (or could have) adverse consequences way beyond our lifetimes. Where crimes are committed as part of a widespread or systematic attack against any civilian population they are considered to be a crime against humanity. The term 'humanity' indicates that the crime is a concern to all of humanity, such is the gravity of the crime that when it is committed, all of humanity is injured and aggrieved. Crimes against future generations are similar, and arise where there is a connection, in terms of knowledge and causation, between the underlying offence and damage to future generations of life. The Bianca Jagger Human Rights Foundation (BJHRF)[2] and Professor Otto Triffterer, Former Dean of the Law Faculty of the University of Salzburg and Editor of the *Commentary on the Rome Statute of the International Criminal Court*, are advocating that the International Criminal Court's jurisdiction should be extended to cover crimes against future generations that are not already proscribed by the ICC's Rome Statute as crimes against humanity, war crimes, or crimes of genocide. The definition of a crime against future generations asks that '*conduct which places the very survival of life at risk should be prohibited and prosecuted as an international crime.*'

Acts which constitute crimes against future generations are acts or conduct committed with knowledge of their '*severe consequences on the health, safety, or means of survival of future generations of humans, or of their threat to the survival of entire species or ecosystems.*' Acts include military, economic, cultural or scientific activities, or the regulatory approval or authorization of such activities, which:

a) cause widespread, long-term and severe damage to the natural environment;
b) gravely or irreparably imperil the health, means of survival or safety of a given human population;
c) gravely or irreparably imperil the conditions of survival of a given species population or ecosystem.

Inclusion of crimes against future generations into the Rome Statute would impose upon member states the duty to investigate, arrest and prosecute perpetrators. Crimes against future generations are not future crimes, nor crimes committed in the future. Rather, they apply to specific acts or conduct undertaken in the present which amount to serious violations of the economic, social and cultural rights of members of any identifiable group or collectivity, or have serious repercussions for the natural environment in the present, and which are substantially likely, as assessed in the present, to have severe consequences on the long-term health, safety, or means of survival of this group or collectivity.

UNIVERSAL DECLARATION OF THE RIGHTS OF MOTHER EARTH

Extending the doctrine of the well-being of life to all life inevitably also includes respecting the rights of nature. In April 2010 Bolivia presented a Universal Declaration of Mother Earth Rights to the people of the world at their World People's Conference on Climate Change[3], in which over 30,000 people participated. A working group addressed Mother Earth Rights, drafting the Declaration of Mother Earth Rights which is to be submitted by Bolivia and other member states to the United Nations. The Declaration will be an important starting point for international engagement on the rights of nature.

Article 1. Mother Earth

1. Mother Earth is a living being.
2. Mother Earth is a unique, indivisible, self-regulating community of interrelated beings that sustains, contains and reproduces all beings.
3. Each being is defined by its relationships as an integral part of Mother Earth.
4. The inherent rights of Mother Earth are inalienable in that they arise from the same source as existence.
5. Mother Earth and all beings are entitled to all the inherent rights recognized in this Declaration without distinction of any kind, such as may be made between organic and inorganic beings, species, origin, use to human beings, or any other status.
6. Just as human beings have human rights, all other beings also have rights which are specific to their species or kind and appropriate for their role and function within the communities within which they exist.
7. The rights of each being are limited by the rights of other beings and any conflict between their rights must be resolved in a way that maintains the integrity, balance and health of Mother Earth.

CLIMATE JUSTICE TRIBUNAL

Working Group 5 of the World People's Conference examined the objective of creating a Climate Justice Tribunal to assume responsibility for the prevention and punishment of climate and environmental crimes that violate the rights of Mother Earth. It is proposed that the Tribunal has the authority to judge, civilly and criminally, states, multilateral organizations, transnational corporations, and any legal persons responsible for aggravating the causes and impacts of climate change and environmental destruction against Mother Earth. Claims may be brought by a civilian, a nation or a community who have been affected.

INTERNATIONAL COURT FOR THE ENVIRONMENT [4]

In June 2008 an early day motion (No.1820) was moved in the UK House of Commons welcoming the proposal to 'establish an International Court for the Environment as the supreme legal authority for settling issues regarding harm to the environment, and for specifying the ecological conditions which must be met if the biosphere is to operate effectively without a major disruption to human communities and other life.' An early day motion (EDM), in the Westminster system, is a motion tabled by members of Parliament for debate 'on an early day' in the House of Commons. However, very few EDMs are actually debated. Instead, they are used for reasons such as publicising the views of individual MPs, drawing attention to specific events or campaigns, and demonstrating the extent of parliamentary support for a particular cause or point of view.

Support for an International Court for the Environment with attendant international rules and laws to be created to govern the planet has grown over the past few years in the UK. Lord Bingham, the former senior law lord in the United Kingdom (2000–2008), speaking on 'The Rule of Law in the International Order' called for internationally agreed and implemented rules if the daunting challenges now facing the world are to be overcome. [5]

Stephen Hockman QC, put the case for an International Court for the Environment at the United Nations Association UK's eighth annual Ruth Steinkraus-Cohen International Law Lecture in March 2010, in which he proposed a body similar to the International Court of Justice (ICJ) in The Hague to be the supreme legal authority on issues regarding the environment. The Court, to be led by retired judges, climate change experts and public figures, would also fine countries or companies that fail to protect endangered species or degrade the natural environment and enforce the '*right to a healthy environment*'. As well as providing resolution between states, the Court could be used to ensure multinational corporations abide by the national environmental laws of all countries with whom they conduct business. The Project on International Courts and Tribunals based at the University College of London is an established body which could undertake research on this proposal.[6] A draft Protocol setting out the 'constitutional rules' of an (admittedly non-binding) ICE now exist.

The idea of an international court that specifically addresses environmental issues has, in the last few decades, been mooted by many and has developed over time. In 1986 an International Forum on Justice and Environment was held in Rome after the Chernobyl disaster. One suggestion was for a Supranational Authority to protect the environment by imposing fines for environmental damage that was considered to be economic damage. Two years later, the idea had developed further. The International Court of the Environment Foundation (ICEF) was established as an NGO, presided over by Amadeo Postiglione, Justice of the Italian Supreme Court.[7] In 1989 the First ICEF Conference was held at a global level. It expressly linked the question of a more effective International Human Right to the Environment to that of creating an International Tribunal for the Environment. In 1991 an international conference was held in order to examine the project for an International Court of the Environment. The initiative was organized and chaired by the Italian Supreme Court and was supported by the Tuscany Region, the Province and the Municipality of Florence, the European Commissioner for the Environment and authoritative experts from 25 countries. From the institutional and moral point of view, there was the support of the Holy See and the Chief Justice of the Italian Constitutional Court.

Soon after, ICEF participated at the 1992 Rio Earth Summit as contributors who were proposing the creation of an International Court of the Environment at a global level. On 9th March 1992 the European Parliament adopted a Resolution on the constitution of an International Tribunal for the Environment.[8] The Resolution was not discussed at the Rio Conference.

ICEF participated in the Fifth Conference on the Mediterranean and Black Sea Basins (1999), organized in Turkey by the Council of Europe, and submitted the project raising the interest also of the Interparliamentary Union and the Biopolitics International Organization (2001) along with the Secretary General of the Permanent Court of Arbitration, members of the Council of Europe and the European Commission.[9]

Moving forward, the Italian Ministry of Foreign Affairs hosted a Global Environmental Governance Conference on 21st May 2010 pulling together many of the parties working on the proposal. The objectives of the ICEF include the promotion of a balanced system of global environmental governance, which includes the political-administrative profiles and the jurisdictional ones; the creation, in particular, of an International Environment Agency, with functions of more effective control, monitoring and management of the environment; and of an International Court of the Environment (whether it be a stand-alone court, ad-hoc courts or an adjunct to the International Criminal Court) giving access not only to states but also to associations and to individuals and the inclusion of international environmental crimes within the competence of the International Court of the Environment or of the already constituted International Criminal Court (ICC).

Global Commons

Commons trusts, where trustees are appointed to undertake legal and fiscal responsibility for the long-term preservation, use or production of a commons, are legal constructs weaving principles of (private) property rights, (public) sovereignty rights, and (common) sustainability rights with a duty of care. The Atmospheric Air Trust is a proposal applying trust law to protect the air's right not to be polluted. The benefit to the atmosphere is tangentially beneficial to humans as well – quite literally by imposing laws which recognize the right of the atmosphere to be clean and our duty to maintain it, we all breathe better.[10]

Increased use of commons trusts to protect global commons will set political priorities for the access to and allocation of global common assets in the 21st century. In the creating of trusteeships, rights to our commons may be realized for the benefit of all. A commons can apply to a natural resource (replenishable or depletable), the value of which is created through the preservation or production of the common goods. Rules governing people's access to – and benefit from – these common resources are for the benefit of a) the people who share this commons (users, managers, producers and providers) and b) the wider community.

Proposals to Protect the Commons

Taking responsibility for our traditional commons to reclaim and restore them involves identification of different types of commons. Full recognition of people's rights to their commons requires a new system of economic exchange in which both streams of common goods – traditional and emerging – are preserved or created independent of commercial and financial pricing. The Global Commons Coalition[11] proposes a system where:

- common goods are protected for future generations;
- some portion of these resources are rented to businesses for the production and consumption of private goods by the present generation; and
- taxes attached to the usage of a commons are redistributed by the state as public goods, to provide an income for those who have been negatively affected by the extraction and production of their common resources, and to repair and restore the depleted commons.

Taxation of land, as a land rent rather than land ownership, is an idea that has been espoused by economists as a far fairer system for centuries. To date few countries have followed this route despite the economic stability it affords, as has been demonstrated by the equal exchange systems in Denmark and Taiwan. Economist Adam Smith in *An Enquiry into the Nature and Causes of the Wealth of Nations* suggested that the best way of funding the state would be a ground rent paid by everyone holding land; the late 19th century economist Henry George was the most influential advocate of funding government by a 'single tax' on land values.

His ideology, known as Georgism, was premised on the principle that everyone owns what they create, but that everything found in nature, most importantly land, belongs equally to all humanity.[12]

Whether our commons are traditional (rivers, forests, indigenous cultures) or emerging (solar energy, creative commons licences, internet), communities that are created under global commons trusts can administer them through unique forms of self-governance, collaboration and collective action. By working together to preserve these resources, new standards of responsibility are established to ensure mutual aid and sustenance for all beings.

The global commons consists of all of these shared resources with fair and equal access. A living trust, or inter vivos trust (Latin for 'between the living'), can be set up. Such a trust is normally created during a person's lifetime to set up long-term property management; in place of the asset being property which will benefit one person, the global commons can be put in trust for the benefit of all people. International law can be used to set up global commons trusts to protect the planet. Using the principles of trusteeship law, we can protect land that is threatened with destruction. Similarly, people can set up a commons trust for their community in which the trustees take on the responsibility to be stewards of the shared resources for the benefit of the wider community and implement a set of governance principles that go beyond mere sustainability to guide their stewardship.

The globalization of political systems and interconnectivity of economies and information networks has created new possibilities for numerous commons partnerships, using principles and linkages that reach from the local levels of social and political organization to higher levels of multilateral governance. It calls for a new framework of global interaction and dialogue based on trust law.

To establish partnerships for stewardship of the global commons is a recognition that the various ecosystems, communities and practices which appear separate are actually part of the same phenomena. Bringing these various relationships together on a global scale will help foster peace.

Yasuní ITT initiative

The Yasuní-ITT initiative demonstrates to the international community how the rights of nature, now embedded in the Ecuador constitution, can be applied. On the 3rd August 2010 Ecuador's President Correa signed the Yasuní ITT initiative to prevent the extraction of oil from the Yasuní National Park. Yasuní is one of the most biodiverse regions on Earth, some areas so remote that there are tribes living there who have never had contact with the outside world. It is also sited on an enormous bed of oil which was discovered a few years ago. The government has agreed to leave the oil underground indefinitely. In oil term, this equates to 850 million barrels of crude oil being left in the ground, preventing 410 million metric tons of carbon dioxide from being released.

To finance part of the project, the public will be able to buy a barrel of oil and leave it underground. The project aims to engage people throughout the world to be part of the change and become defenders of the rights of nature by protecting the Ecuadorian Amazon jungle and the interests of indigenous groups living in voluntary isolation in the Yasuní National Park.

Speaking from the Amazon, one leader explained "We never gave up hope and our wishes have come true. The land has been spared and all that live there can breath again. The government has answered the pleas of the Earth, the Tagari and Taromenane people who live in voluntary isolation and all the environmentalists who have been involved in raising the profile of Yasuní."

Value the planet and one becomes morally disposed to protect it; treat it as a commodity and one has no sense of moral obligation. Now for the first time the rights of nature, the rights of the people who live there and the rights of future generations have finally trumped the silent rights of commerce.

Part 4

TOWARDS A
LIVING PLANET

This law of nature, being co-eval with mankind and dictated by God himself, is of course superior in obligation to any other. It is binding over all the globe, in all countries, and at all times: no human laws are of any validity, if contrary to this; and such of them as are valid derive all their force, and all their authority, mediately or immediately, from this original.

SIR WILLIAM BLACKSTONE,
English judge and jurist, Natural Law Theory, 1761

Chapter 10

THE COMMANDING VOICE OF THE PEOPLE

CORPORATIONS are the ones gambling our planet away and our governments are running the casino. They are taking huge risks on our planet's future. It is a moment of precarious balance – the globe is spinning in the opposite direction to the wheel: winner or loser, our fate lies in where it lands. There is only one chance of winning. Governments can change the rules mid-spin and the ball can be caught and placed peacefully on the winning green. Governments can change the rule that says the world is owned by corporations – faceless pieces of paper that have the power to destroy. Rebalance the scales of justice by putting the health and well-being of people and planet first: establish laws to support trustees, guardians and stewards, to assist the stewards to take action when required, to assist the trustees who are acting on behalf of the people's sacred trust, to enable guardians to speak on behalf of the planet when help is needed. It is our heads of state that can take the crime of ecocide to the United Nations, for it to be made into international law. They are the ones who sit on the world's most powerful committee and who can decide the fate of the planet.

Public engagement is the lever that will turn this game into a different type of game altogether. One that is premised on creativity and collaboration; a game where we all benefit as players who are participating in the process. Public participation is the key to unlocking the door of the game-store; once in we can cherry-pick what we like. Games of innovation, new ventures; games of peace, new alliances. Games to build new communities, joint undertakings for land use, projects which bring quality of life to all who play. We shall have new drawing crayons, of all colours, shapes and sizes to provide the golden seeds to be planted and to illustrate the pictures of the blossoming landscape. The paper used to draw on will be made of constantly renewing matter, changing as required. We shall have new toys, to provide pleasure and benefit, toys which can be shared and used by many. Payment is the promise to play.

Once we have left the game-store we can begin straight away. There is much to do: the new rules require explanation and new systems put in place. Help is available for all who request it and tools shall be supplied to all who ask. Training for players who are unfamiliar with the game shall be provided. Everyone is included. Everyone has a role to play in the new game of life.

Our representatives at the top table are our heads of state. They speak for us. We have voted them in, we have given them their job. To be a member of a parliament is to be in service to society at large. To represent constituents interests is to put the sacred duty of the health and well-being of inhabitants first. Here is how we can embed a very different outcome instead:

MAKE ECOCIDE A CRIME

Break the cycle of harm to wildlife, nature and the land; and in Alberta, contaminated wetlands and ancient forests would be restored. The use of air cannons would be outlawed from being used as deterrents to wildlife. Guidelines would be created for the testing of pesticides based on the precautionary principle, determined by non-industry representatives; all pesticides that are toxic would be banned.

PUT IN PLACE ALL EIGHT PRINCIPLES

Laws to Prevent Ecocide
1. Amend all compromise treaties, laws, rules and regulations:
    - (i) replace with prohibition of all damaging and destructive practices; and
    - (ii) include provisions to enable restoration of damaged territories to be prioritized over existing practices that are premised on financial penalty alone;

Sacred Trust of Civilization
2. Community interests to be placed over private and corporate decisions;

Holding Business to Account
3. An elected community member of the territory concerned to be seated on a company's board of directors;
4. Accountability of business practices to be scrutinised by independent bodies and made publicly available; transparency of process to be open to public comment;
5. Where lobbying at a political level is undertaken by corporate entities, a record of all activities and monies spent to be made available for public scrutiny and query;

Environmental Sustainability
6. Hold all governing bodies to account. If damaging policies are being promoted, call for them to be halted with immediate effect;
7. Hold community meetings and ensure proper democratic and true consultation is seen to be done;
8. Transition to cleaner solutions to be rapid and effective.

GOVERNMENT CHECKLIST
Make ecocide an international and national crime; imposing a duty of care to assist those affected by mass destruction, damage to or loss of ecosystems, holding to account the persons who

have superior responsibility for the harm caused and impose the burden of restoration on the company, at whatever cost that may require. Implement Rules of Procedure for Environmental Cases, including: protection from SLAPPs, the Precautionary Principle, The Writ of Kalikasan and Environmental Protection Orders.

Make Category A banking projects illegal and prioritize Category C projects. Ensure the sacred trust of civilization is upheld; that the interests of the inhabitants of a given territory who have not yet attained a full measure of self-government are paramount, and the obligation to promote to the utmost the well-being of the inhabitants of these territories. Ensure the peoples' right to self-determination is respected in all decision-making; and develop friendly relations among nations based on respect for the principle of equal rights and self-determination of peoples. Uphold the equal rights of both men and women. Crimes committed against future generations are happening today and have consequences for the unborn humans of a future time: create Crimes Against Future Generations. Vote for and call for the Universal Declaration of Mother Earth Rights to be recognized throughout the world. Put in place an International Court for the Environment and an International Environment Agency, with functions of more effective control, monitoring and management of the environment. Create land value taxation attached to usage of land, to be redistributed by the state as public goods, to provide an income for those who have been negatively affected by the extraction and production of their common resources, and to repair and restore the depleted commons: ensure the global commons are shared resources with fair and equal access. Call for international law to be used to set up a global commons trust to protect the planet and mandate that all NGOs and indigenous community organizations have 'participant status' at all world summits and negotiations. Ratify the Convention on Civil Liability for Damage Resulting from Activities Dangerous to the Environment and make sure it is individuals who are held to account.

Halt the discharge of toxic substances and the release of heat in order to ensure that serious or irreversible damage is not inflicted upon ecosystems; impose strictly liability on all operators of installations for all damages caused by engaging in dangerous activities. Put into operation the EEC Directive on the Equal Treatment of Men and Women, including at board level. Impose a 1% tax on business annual revenues, to be paid to environmental organizations. Improve the international competitiveness of trade and industry by putting in place a Financial Statements Act (accounting for CSR in large businesses), making it mandatory for all private companies, investors and state owned companies to include information on CSR in their annual financial reports. Ensure all businesses provide mentoring schemes.

Businesses can change their investment strategies. For BP, going beyond petroleum was a possibility, but what was missing was the support of law. Their voluntary action in response to their own growing concerns about climate change and damage to the environment could not happen in isolation. This time around, ecocide law can ensure that we do go beyond, and fast.

If you have a pension, you can contact your pension provider and ask questions as to what and where your money is invested in. Pension schemes are invested in up to a fifth of British companies. Even though pension schemes can have so much leverage over what a company does – changing their attitude to climate change, child labour or arms sales – these issues are rarely raised with the companies invested in. It only takes a moment of your time to request your pension scheme to become a responsible investor, asking them to be environmentally, socially and financially responsible. Seen another way, this is about taking responsibility for the way our capital is being used to shape our world that we and our children live in.

If your money is tied up in dirty and polluting businesses, let your provider know you are not happy with that and that you'd like your pension invested in clean business. If there is no green or ethical funding alternative, ask why not; such alternatives do exist.

You may wish to switch to a fund manager or pension provider who provides non-harmful investment. If you have a bank account, you can find out what your bank is using your money for. Find out what their ethical and environmental policy is and compare it to the Co-operative bank. If it doesn't stack up, change your bank and let your manager and the head of the bank know why. It may sound inconsequential, but it is not; you can do the same with your investment provider. A couple of minutes spent googling 'ethical investment' opens up a whole new world of what is possible. You can put your money into what you choose the future to be, after all it's your money. If you have shares, you can take a far closer look to see what sort of business you are propping up, and if it doesn't stack up, start asking questions, go to the company's AGM and ask questions there and see what response you get. You have a right to be heard. If you do not like what you are hearing, get in touch with FairPensions and see if there is a resolution that you and other shareholders can submit. Do not underestimate the power you have as a consumer, especially when you let it be known why you have decided not to support your bank/pension provider/fund manager.

People throughout the world over the centuries have refused to accept wrongdoings that occured against them, their home-land, their families, children; decisions that precluded them because they were a woman, the wrong colour, from the wrong background, for having a disability. Time and again it takes concerted efforts to speak up for justice when it goes against prevailing beliefs and sometimes it takes more than one generation to succeed. Often we know not when the tipping point shall be. Those that have preceded us have rallied hard so we can take for granted that which we have today. We in turn have new challenges, in different arenas with different issues, but in reality they often come down to the same common denominator. Calls for public and environmental justice are about securing peaceful life for future generations, the aspiration for a better world. Often so few live to enjoy the fruits of their newly found freedoms. To me, this is the ultimate altruistic act; to give your life in service for a greater good. My rallying call is for life for all who inhabit this planet, not just humans. It can be your call too if you so choose. I ask you to join me.

Reading about the life of William Wilberforce some years ago had an enormous impact on me, to see how history was repeating itself and how he, along with just a very small group of people stood up and commanded justice tirelessly, stopping at nothing until he succeeded. Wilberforce vowed back in 1787 that slavery was 'so enormous, so dreadful and so irremediable that I resolved I would not rest until I had effected its abolition.' He gave his whole life to achieve what he set out to do – and succeeded. If that's how long it will take for us to effect the abolition of ecocide, then that is what I committed to doing quite some years ago. I like to think that in this day of internet and 24 hour access to virtually anyone anywhere in the world that it may just take a good deal less time.

Much of the work for this book has come together in a very short timespan, built on seven years of examining the what, why and how of law and how we can use law creatively to protect our *oikos*. My experience of being in Copenhagen at the collapse of the climate negotiations in December 2009 was an enormous sense of pain, despite knowing that it was in some respects a good outcome. I sat in pent up frustration with good friends watching the scene unfold at the Bella Centre, where many heads of state or their representatives were based just a mile along the road from where I was, at the People's Climate Summit. Much as there was an enormous desire to have a document of some sort, in the end very little could be agreed. Crucially, what was agreed was the acceptance that prevention of dangerous anthropogenic interference is required.

Nature sometimes generates her own forest fires to clear away dead wood, to make way for new growth. And so it was with the Copenhagen climate negotiations. To build a construct to protect the planet on market mechanisms is to build on sand. The negotiations had to collapse. In doing so, it opened the door to the possibility of new ways of thinking and working this problem out. This is one arena that should not be open to negotiation.

Copenhagen was the catalyst that strengthened my resolve to find a better way forward. It took three months of concerted single-minded thought and research to conclude that an international crime of ecocide could work, but more than that, is absolutely necessary if we are to turn this sinking ship around dramatically. My proposal was submitted to the UN in April 2010. So much has happened since then; and yet, the international political arena has still to take the next step – preventing dangerous anthropogenic interference means bringing to an end dangerous industrial activity. It's a big ask but the enormity of the issue at hand requires nothing less. It has been tried before, I have discovered. Amazingly the crime of ecocide has a history of almost becoming an international crime stretching back to 1945. In 1972 Olof Palme, the ex-prime minister of Sweden, opened the Stockholm Conference with a call for law to stop ecocide, a convention was drafted the following year, and ecocide was included in the Rome Statute text until it was removed in 1996, at the 11th hour.

Writing this book has been a journey in learning to harness my frustration, anger and pain. I have had to learn how to channel my emotions constructively, making sense of the wanton and reckless destruction that lies scattered across the planet, quite literally as debris on an ever-expanding wasteland. I was challenged to examine what I really valued in my own life, to apply to my own life the principles which I propose for the wider world. I have sat and howled at the injustice upon witnessing sights of enormous devastation. Such as been the pain that I thought I would be seared. If you have got as far as this reading this book, then I feel less exposed in admitting this – you too care. I have learned not to rage at faceless corporations but to see them as communities of people, who often want the best for their families and homeland too but have no idea how to get there or to stand up amidst others who are shackled to a system that pushes profit above all other considerations. Laws that do not work have created that paradigm; directors have a legal duty to ensure profit is their number one priority. So few dare to challenge this premise. Different, new radical laws such as the ones suggested here can easily change the framework in which we function. Of course laws in themselves are not enough, it is the governance systems that are put in place that provide the necessary support.

Indigenous people talk about the importance of witnessing, and I think maybe this is key – once I saw, really saw, and connected at a very deep level to the inherent wrongness, only then could I move on and explore what had to be done to bring about change. To turn away and refuse to face the reality is to remain numb. It is also a state of being – of being passively complicit. I chose to see.

Each chapter brought with it a deeper understanding, a deeper commitment and resolve to see the principles I have proposed put in place. They are simple principles and I believe they can be applied across the board, in the smallest of communities to the largest of intergovernmental decision-making arenas. Many of the laws and governance systems we could use to apply the principles are already in existence, some of which I did not know of until I looked for evidence elsewhere in other countries. As I wrote and researched I woke up to the fact that law is still required – without it a relatively small number of individuals, who are in positions of superior responsibility, evade their overriding duty of care. I think we can change how we make our decisions: in setting out what matters, first and foremost, it is far easier to discern what laws, policy and strategy we require to bring it to an end. To me what matters most of all is the well-being of our *oikos* – our home and all who inhabit this world. BP is just one example of a corporation, like many others, where profit comes first. There is very rarely any intent – the intent is to maximise profits, but whether or not there is knowledge matters. The starting point is whether significant harm is caused *per se* – if it is known there could be significant harm further down the line, then it becomes an aggravating feature. In legal terms, what I speak of here goes to sentence only.

Researching our recent legal past, for the clues as to how we have ended up in the precarious position we now find ourselves, was an eye-opener for me. So much in our lives has a connection one way or another with ecocide, if we look further down the chain and more closely examine how we source much of what we take for granted. We are all in some way implicated. Ecocide does not just happen on an enormous scale, over there in some distant continent, but also at far smaller levels too. None of us come to the table with clean hands.

This book was written in just under ten weeks in support of my written legal submission on Ecocide law to the United Nations, which sits at the heart of this book. Everything else in my life was put on hold. Ten weeks of concentrated unravelling of the spaghetti of jumbled ideas, pulling out a strand and carefully holding it up to the light then putting it into words, giving context by adding true stories to illustrate my arguments. I returned to first principles, researched primary texts and extrapolated the essence. Many of the proposals are spoon-sized, for others to explore further and implement. Please, take these proposals and make them your own. Nobody has ownership over ideas. Least of all, an idea whose time has come.

I hope I awaken in lawyers and law students the power of the law and how it can be used constructively, to protect the greater good. If I inspire even one person to take action as a voice for the Earth, and if that person is you, know that you shall inspire many others after you too. Discerning what makes for a law that works (putting the interests of people and planet first) and what is compromise law, can solve so many problems from the outset by preventing laws from being put in place that are not fit for purpose. Now more lawyers are joining the activists who are already taking a stand for what they believe is causing harm and has to be stopped. The world of business is waking up too; some directors and CEOs are speaking out in the boardroom. The next step is for our leaders to stand up for the Earth too, to be fearless and act from a place of deep care.

I seek peace. I seek a world where we no longer have laws, because societies are capable of self-determining collectively for the greater good, so that the court door remains closed and the keys are eventually handed over to the community who can then put the building to another, better, use. We're not there yet, it may take a few generations. This book provides bridges to that world. You may think this is a complete utopian dream - but why not dream for the best? Aspire for nothing less. If we can hold a vision of what that could look like then we have the potential within us to bring it into being.

One of the major bridges to take us to a better world is Ecocide law. There is a misalignment here – only when human law is aligned with higher law shall harmony prevail. So if you like, a realignment of our human law is inevitable when we seek harmony instead of harm. I invite you to take this path. Pass this book along, gift it to others to read, request your local independent bookshop and library to stock it, seed it to your friends and family – to anyone who may care, seed it wherever and to whomever it feels best. This is one idea whose time has come, and with your help it can really spread. Teachers and parents, you can – if you so wish – help empower the next generation. Together we can uproot the old, whilst seeding and nurturing the new.

I ask you, come, stand up and be counted. Come with me and command that this ecocide stop. This is our right: it is also our responsibility. Together our voice can be all-powerful. Together we can make this world the one we want – and attain the freedom of life itself.

BP's Reliance on Global Inaction

The International Energy Agency (IEA) is an intergovernmental organization that was established in the framework of the Organization for Economic Co-operation and Development (OECD) in the wake of the 1973 oil crisis. The IEA serves as an information source on statistics about the international oil market, which includes responding to physical disruptions in the supply of oil. Every year the IEA publishes its World Energy Outlook Report and gives scenario projections, which many governments rely on. Against its 2009 Reference Scenario,[1] which is the 'business as usual' scenario, various possible scenarios are compared. The Reference Scenario, which assumes no change in government policies by 2030, sees world primary energy demand a dramatic 40% higher than in 2007 and up to 80% being met by fossil fuels. By way of comparison, the '450 ppm Scenario' (reducing greenhouse gas emissions to a 2 degree temperature increase) is 34% less than in the Reference Scenario and offers, say the IEA, "important energy security and environmental co-benefits."

On 5th March 2010 BP placed on its website a document entitled *Oil sands resolution and response*, a four page reasoning as to why BP opposed the proposed Shareholder Resolution which had been raised by some of their shareholder investors calling for disclosure and accountability of BP's recent change in policy to invest in unconventional oil extraction from the Canadian Tar Sands and in particular their investment in the Sunrise/Toledo project. BP argued that their energy strategy was an appropriate and correct strategy to advance on the basis that 'World energy demand is projected to increase by around 40% between 2007 and 2030 with fossil fuels still satisfying as much as 80% of that demand by the end of that period.' The IEA Factsheet was footnoted in support of their reliance of these figures. [2]

In a letter sent to BP the 18th of March 2010, FairPensions (the NGO which co-ordinated the filing of the resolution) and various co-filers raised their concern that BP was reliant on a demand projection taken from the IEA's Reference Scenario. FairPensions pointed out that this was a scenario which the IEA had admitted is 'a baseline picture of how global energy markets would evolve if governments make no changes to their existing policies and measures.' By BP's own admission, they were intent on pressing forward on the basis of a future as set out in the IEA Reference Scenario.

The FairPensions letter challenged BP's reliance on the IEA Reference Scenario. To do so, explained the letter, leads to the conclusion that BP "assumes that there will be no governmental action on climate change and that a six degree temperature increase will occur resulting in catastrophic climate change. In our view it is not feasible to base business strategy on a scenario which the international community is committed to ensuring never comes to pass." In light of their confirmed intent to achieve 80% (use of fossil fuel in the energy mix by 2030), the letter queried BP's reliance on this demand projection given their own admission that "it is likely that policy responses to climate change, energy security and energy poverty will profoundly affect future outcomes."[3] Despite their stated knowledge of the likelihood of policy restrictions being imposed, BP were nevertheless intent on proceeding with a strategy premised on six degree temperature increase and were not prepared to examine and change their own policies.

On the 15th April 2010 BP held their AGM in London. Fair-Pensions succeeded in securing a 15% special shareholder resolution vote against BP management despite BP's fierce opposition. That is a strong vote for a resolution of this kind. It was a remarkable attempt by institutional investors to exercise stewardship over their shareholdings in order to protect the interests both of their end-beneficiaries and of the wider public. The Chairman Carl-Henric Svanberg, in delivering BP's formal response to the shareholder resolution, confidently confirmed that BP were working on the premise that demand will increase by 40% between 2007 and 2030 with fossil fuels/hydrocarbons meeting up to 80% of that demand. Louise Rouse, FairPensions Director of Investor Engagement, stood up to ask the Chairman: "given that BP agreed with the demand figures in the Reference Scenario, did BP also agree with the assumptions underlying that scenario i.e. that there would be no government action on climate change and that there would be a six degree global temperature increase?" The question she asked was not answered.

Epilogue

A N epilogue normally reveals the aftermath of the people and their stories. Many who are named in this book speak on behalf of countless others who stand strong beside them, calling for justice. For the people of the Dongria Kondh, their voice was heard. In 2013, a landmark ruling from India's Supreme Court upheld the Dongria Kondh's right to save their sacred hills. Vedanta, it ruled, had no say. The Dongria were to decide whether to allow mining on the Mountain of the Law. Naturally they said 'no'.

The real-life character in this story, the global story of eradicating ecocide, is you. What you make of this book and what you decide to do next is what will determine our future. It is our voices that can inspire our leaders to sound the horn. Yes, of course we can embrace the proposed governance and laws that are included in this book; and yes, to do this shall take great courage. We have a wonderful chance to change our outcome. What happens to our planet is up to us: only we can choose to stop the mass ecocide, for present and future generations. To amend the Rome Statute to include ecocide requires just one head of state to launch the process. This is emergent law and whether or not it is put in place comes down to at least one, and preferably more, state calling for it. I am just a lawyer proffering a legal route-map forward. Whether or not it is taken forward, is an echo of whether or not civilisation as a whole is ready to take that quantum leap.

There are numerous individuals and organizations who are actively involved in changing the systems around us. Many have influenced my journey, expanding my vision as I have searched to discover the trim tabs. I give thanks for the invaluable contributions which have prompted me to explore avenues I would not otherwise have taken. I applaud the work so many are doing to ensure we leave this world a better place than when we arrived. The most powerful voices of all are those who are often working in small groups, decentralized networks which are collaborative in nature. Here are some people who are the voices for their causes and the pathways to their stories:

Allan Savory set up the Africa Centre for Holistic Management turning deserts into thriving grasslands, restoring biodiversity, bringing streams, rivers and water sources back to life, combating poverty and hunger, all essential interventions to mitigate against global climate change. *Holistic Management: A New Decision Making Framework* by Allan Savory and his wife Jody Butterfield is an essential handbook for anyone involved with land management and stewardship – ranchers, farmers, resource managers, and others – and a valuable guide for all those seeking to make better decisions within their organizations or in any aspect of their personal lives. See *www.savoryinstitute.com*. The story of the restoration of the Loess Plateau has been filmed and can be viewed at *www.earthshope.org*.

Louise Rouse of FairPensions, who challenged BP's strategy of expansion based on their knowledge of climate scenarios, continues to challenge unethical institutional investments and call for scrutiny of pension investment. To challenge where your pension is invested, go to *www.fairpensions.org.uk/getinvolved*.

Rights of other beings have been honoured by numerous communities in various traditions throughout the world for countless years, based on life supporting values. **Cormac Cullinan** is a leading advocate of new laws to protect the natural world. His seminal book, *Wild Law – A Manifesto for Earth Justice*, sets out the reasoning why earth governance is rooted in transforming human perception of nature into a living daily practice of deepening our connection with the whole.

Thomas Berry, philosopher, cultural historian, geologian and author, was a tireless spokesman for a new creation story. He sought to shift the focus of religion from individual salvation to care of the Earth through appreciation rather than exploitation of the world around us. His writings, which include *The Great Work, Dream of the Earth* and *Evening Thoughts: Reflecting on Earth as Sacred Community,* examine the essential task of the present generation to transition into an ecological era.

Various individuals have widened the field of earth jurisprudence, in particular: **Ed Posey** and **Liz Hoskins** at Gaia Foundation: *www.gaiafoundation.org,* **Pat Siemens** of the Centre for Earth Jurisprudence: *www.earthjuris.org* and **Michel-Maxime Egger** of Fondation Diagonale: *www.fondationdiagonale.org.*

Eriel Tchekwie Deranger is a Dene woman from the Athabasca Chipewyan First Nation of Northern Alberta. I first heard her speak about the tar sands at the Indigenous Peoples' Global Summit on Climate Change in Anchorage, *www.indigenoussummit.com.* Her words and pictures brought home the enormity of the destruction happening in Alberta, Canada. *www.ienearth.org.*

Sue Dhaliwal has single-mindedly and eloquently raised the UK profile of the injustices of unconventional tar sand extraction, one of the largest and unrecognized ecocides in the world. Sue stood up when others would not: *www.no-tar-sands.org.* Platform's excellent report on the tar sands, *Cashing in on Tar Sands,* can be downloaded at *www.platformlondon.org.*

Ana Simeon and all at Sierra Club BC: without their private prosecution, the plight of the ducks and the toxic tailing ponds would never have come to the public's attention. *www.sierraclub.ca/en/tar-sands.*

An incredible amount of direct action has mobilised in the UK and elsewhere – especially those faced with their own ecocides. In Australia the **Lock the Gate** movement continues unabated, in the USA, UK and other EU countries **Frack Free** communities build every day, as do other grass-root movements such as **Via Campesina** and **Idle no More** – all are compassionate revolutionaries.

Isabel Carlisle, whose work is now pioneering alternative education for young people. Alternative education systems can be found across the world (e.g. YIP in Sweden, UnCollege movement), and are fast growing to meet the needs of a new world. Isabel is the founder of *School in Transition*, and the one-year skills-based *Transition Learning Journey* for young adults who seek a learning programme outside the college/university system. Part of the TLJ programme are 3-4 month placements in Transition Town Initiatives.

Misty Oldland, a font of creativity, is the magic behind The Golden Seed. It is the story of a heroic group of endangered species and their quest to save the planet. *www.thegoldenseed.co.uk*.

Roz Savage is an active environmental campaigner and adventurer with a difference; she rows the oceans to bring attention to the devastation wrought on our seas. A woman of remarkable tenacity, what she has to say in her book is powerful: *Rowing the Atlantic – Lessons Learned on the Open Ocean.* See *www.rozsavage.com*.

Dr. Helen Caldicott is the single most articulate and passionate advocate of citizen action to remedy the nuclear and environmental crises. She has written outstanding books, convincingly presenting the case against nuclear power. Her books include *War in Heaven, Nuclear Power is not the Answer* and *The New Nuclear Danger.* See *www.helencaldicott.com*.

Atmospheric Trust Litigation was pioneered by **Professor Mary Wood**. Since 2011, youths from across the USA have brought litigation against 50 US states holding to account and asserting a legal duty of care under the International Trust Campaign. You can read more at *www.ourchildrenstrust.org*.

Tamsin Omond speaks for the planet in everything she does, from standing up against aviation expansion to calling on politicians whilst dressed as a suffragette. *www.climaterush.com, www.planestupid.com*.

My meeting with **Joanna Macy** opened a new door, of exploring my inner ecology, which proved to be a definitive turning point. Her most recent book is *World as Lover, World as Self: A Guide to Living Fully in Turbulent Times.* *www.joannamacy.net*.

Georgina Downs has campaigned since 2001 for regulatory

changes. The Royal Commission on Environmental Pollution 2004 report agreed with her claim that current regulations are inadequate. The report proposed a five-metre buffer zone be imposed around any agricultural land that is subject to spraying. This small buffer zone is an unacceptable compromise. Georgina continues to fight for better protection for people exposed to pesticides in agricultural areas. To find out more about her work, visit Georgina's website: *www.pesticidescampaign.co.uk.*

I wish also to mention Isabelle Bratt, whose master thesis on the protection of the natural environment during international armed conflict proved to be an invaluable piece of research. The Philippines Environmental Rules and Procedures were sent to me by Chris Milton, to whom I am grateful. Anthony Werner, my publisher, for believing that I had something worthwhile to say; Emma Jell, who was my fantastic PA at the time of writing this book and Victoria Hands, a tour de force and kindred spirit. Thank you all for your invaluable input.

This is a time for courage and leadership both in small communities and large; a time for constructive and timely intervention. No more excuses, no more postponements, no more procrastination. We can speak for the Earth – we are the Earth. Our fundamental imperative is first to do no harm. We have an obligation to restore the lost world that our civilization has destroyed. It is our sacred trust; it is our sacred duty, a duty which we can choose to uphold. Eradicate the ecocide and all life on Earth will once again flourish.

Notes

INTRODUCTION

1. Cain Burdeau and Holbrook Mohr, *BP Didn't Plan for Oil Shock*, ABC News, 30th April 2010: "In its 2009 exploration plan and environmental impact analysis for the well, BP suggested it was unlikely, or virtually impossible, for an accident to occur that would lead to a giant crude oil spill and serious damage to beaches, fish and mammals." www.abcnews.go.com/Business/wireStory?id=10515973

2. James Randerson, *UK's Ex-Science Chief Predicts Century of 'Resource' Wars*, The Guardian, 13th February 2009, www.guardian.co.uk/environment/2009/feb/13/resource-wars-david-king

3. Marla Cone, *Silent Snow: the Slow Poisoning of the Arctic*, Grove Press 2005: a superb investigation into the Arctic's chemical crisis and the tragedy of the contamination of one of the most beautiful wildernesses, and its people, in the world.

CHAPTER 1: TAKING STOCK

1. There is one notable exception: The United Nations Convention on the Law of the Sea (UNCLOS) concluded in 1982, came into force in 1994. To date, 158 countries are signatories. It is now regarded as a codification of the customary international law of the seas.

2. Simon Winchester, *The Map that Changed the World: A Tale of Rocks, Ruin and Redemption*, Penguin 2001, pp. 48–49

3. Peter Thorsheim, *Inventing Pollution: Coal, Smoke, and Culture in Britain Since 1800*, Ohio University Press 2006, p. 5

4. Robert Multhauf, *Sal Ammoniac: A Case History in Industrialization*, Technology and Culture, Vol 6, No. 4, (Autumn 1956), pp. 569–586

5. It was finally repealed and replaced by the Environmental Protection Act 1990

6. William Stanley Jevons, *The Coal Question: An Inquiry Concerning the Progress of the Nation, and the Probable Exhaustion of Our Coal-Mines*, Macmillan & Co 1865

7. Thorsheim, *Ibid*, p. 196 – an excellent analysis

8. David Stradling & Peter Thorsheim, *The Smoke of Great Cities: British and American Efforts to Control Air Pollution, 1860–1914*, Environmental History, 4 (1999) p. 10

9. *Ibid*, p. 11

10. Dr Chris Cook & Dr John Stevenson, *The Longman handbook of Modern British History, 1714–2001*, Longman 2001, p. 257

11. Thorsheim, *Ibid*, p. 198

Chapter 2: Massacre of the Innocents

1. Professor Bill Kovarik, www.radford.edu/wkovarik/misc/blog/8.whaleoil.html. See also Prof. Kovarik's online book, *The Summer Spirit, Myths about Renewable Energy* www.radford.edu/~wkovarik/envhist/RenHist/intro2.html

2. As above

3. US Treasury fact sheet on Civil War Taxes: www.ustreas.gov/education/fact-sheets/taxes/ustax.shtml

4. Robert W McDaniel, *Patillo Higgins and the Search for Texas Oil*, Texas A& M University Press 1989

5. McDaniel, *Ibid*

6. 18 United States Code (1952) § 610. For list of states see G. Kirkos, *Corporations: Political Activities: Interpretation of Statute Prohibiting Political Contributions by Corporations*, Michigan Law Review, Vol. 54, No. 5 (Mar., 1956), p. 701

7. William O Douglas, *Stare Decisis*, (1949) Columbia Law Review 49 (6): pp. 735–758

8. U.S. Department of Defense, Office of the Special Assistant for Gulf War Illnesses. *Environmental Exposure Report: Depleted Uranium in the Gulf (II)*. Washington, D.C. 13th December 2000. See also: Research Advisory Committee on Gulf War Veterans' Illnesses *Gulf War Illness and the Health of Gulf War Veterans: Scientific Findings and Recommendations,* Washington, D.C.: U.S. Government Printing Office, November 2008 ('The Illness Report') p. 86

9. Martin Williams, *First Award for Depleted Uranium Poisoning Claim*, The Herald, February 2004, http://vitw.org/archives/405

10. The Illness Report, *Ibid*

11. John Pilger reporting the effects of DU on the civilian population of Iraq after the 1991 Gulf War: www.pilger.carlton.com/iraq/rokke and www.indymedia.org.uk/en/2004/02/284987.html

12. *MoD Rejecting Gulf War Syndrome Pension Claims, Say Veterans,* Guardian, 11th April, 2010 www.guardian.co.uk/uk/2010/apr/11/gulf-war-syndrome-veterans-pensions

13. The Illness Report, *Ibid,* pp. 311–315

14. U.S. Department of Defense, Office of the Special Assistant to the Under Secretary of Defense for Gulf War Illnesses, Medical Readiness and Military Deployments, *Environmental Exposure Report: Particulate Matter*, 15th October 2002 and The Illness Report, *Ibid,* p. 212

15. A UK Department of Defence map provided to the Presidential Special Oversight Board in 1998 indicated that the highest concentration of DU munitions were fired in southern Iraq and Eastern Kuwait. See: Presidential Special Oversight Board for Department of Defense Investigations of Gulf War Chemical and Biological Incidents. *Final Report*, Washington, DC 20th December 2000. See also the Illness Report, *Ibid*, p. 87 for graph on early research into health effects of DU exposure.

16. U.S. Department of Defense, Office of the Special Assistant for Gulf War Illnesses. (now renamed: Force Health Protection & Readiness Policy & Programs) *Environmental Exposure Report: Depleted Uranium in the Gulf (II)*. Washington, D.C. 13th December 2000, and the Illness Report, *Ibid*, p. 87

17. The Illness Report, *Ibid,* p. 99

18. On 11th May 2009 the Belgian Conventional Arms Law amendments came into force to include prohibition of weapons systems containing DU, dismantling of existing stocks of inert ammunition and armour containing depleted uranium or any other industrially manufactured uranium.

19. www.miningwatch.ca/en/uranium-overview

20. World Nuclear Association, World Uranium Mining, www.world-nuclear.org/info/inf23.html, updated May 2010

21. The Guardian have a nifty DU interactive guide and Special Report explaining the use of DU in Kosovo and Iraq. See: www.guardian.co.uk/uranium and www.guardian.co.uk/uranium/flash/0,7365,420455,00.html

22. DUOB Final Report, p. 52, www.bandepleteduranium.org/en/a/250.html

23. International Institute of Concern for Public Health, *Burning radioactive waste: Blind River and the CNSC decision*, IICPH

Newsletter, 1st October 2007, www.iicph.org/news_0702-blind-river

24.	European Parliament recommendation (2010/2020(INI)), www.bit.ly/bmzn0o For full information and updates of progress: www.bandepleteduranium.org/en/a/335.html

25.	www.thirdworldtraveler.com/Corporate_Welfare/Nuclear_Subsidies.html

26.	More conventional weaponry, such as US B-52 bombers were also deployed, dropping 13 million tones of bombs – triple the tonnage used in World War II. A formation of B-52s could obliterate a "box" approximately five-eights of a mile wide by 2 miles long, striking without warning from 30,000 feet turning entire villages into sudden eruptions of flaming sticks, human limbs and thatch. See Franz J Boswimmer, *Ecocide, A Short History of the Mass Extinction of Species*, Pluto Press 2002, p. 76 for fuller details.

27.	Oklahoma Geological Survey on www.earthquakes.ok.gov. See also www.richardheinberg.com and Richard Heinberg's book, *Snake Oil* (2013) for a comprehensive demystifying and myth busting analysis.

28.	Allegra Stratton, *New Banking Rules Should Reveal Emissions From Investment, Campaigners Say*, 2nd March 2009: www.guardian.co.uk/environment/2009/mar/02/rbs-environmental-regulations

29.	ConocoPhillips loan from RBS, dated 1st October 2008, *Cashing in on Tar Sands: RBS, UK Banks and Canada's "blood oil"*, 2010, p. 52 www.platformlondon.org

30.	*Force RBS to Reform Letter*, published in The Guardian, 29th November 2009: www.guardian.co.uk/theobserver/2009/nov/29/letters-methadone-prison-weather-muslims

31.	See the Financial Reporting Council for full details of voluntary UK Stewardship Code at www.frc.org.uk

32.	See *Cashing in on Tar Sands, Ibid* www.platformlondon.org

CHAPTER 3: TELLING THE TRUTH ABOUT THE BIRDS AND
THE BEES

1.   Natural Resources Defense Council: www.nrdc.org/wildlife/
      borealbirds.asp

2.   www.bancannons.tripod.com/cannons.html

3.   Which accounts for over half the total 221 billion gallons
      created in 2010. See: *Tar Sands Oil Trial Underway – Charge:
      the Death of 1,600 Ducks*, Susan Casey-Lefkowitz's blog, 2nd
      March 2010 http://switchboard.nrdc.org/blogs/sclefkowitz/
      tar_sands_oil_trial_underway_c.html

4.   www.tarsandswatch.org/syncrude-duck-death-trial-
      underway

5.   www.alberta.ca/home/NewsFrame.cfm?ReleaseID=/
      acn/200902/252585BDAB379-DBC4-B789-
      3B6A716375E3BDEF.html, www.environment.alberta.
      ca/02271.html

6.   www.panna.org/legacy/gpc/gpc_200404.14.1.08.dv.html

7.   www.pan-uk.org

8.   PAN Bad Actors are chemicals that are one or more of the
      following: highly acutely toxic, cholinesterase inhibitor, known/
      probable carcinogen, known groundwater pollutant or known
      reproductive or developmental toxicant. Note, because there are
      no authoritative lists of Endocrine Disrupting (ED) chemicals,
      EDs are not yet considered PAN Bad Actor chemicals.

9.   www.chem-tox.com/malathion/research

10.   Parliamentary Papers, 1949, Cmd 7664

11.   Quoted by James Fergusson in *The Vitamin Wars: Who Killed
      Healthy Eating in Britain?*, Portobello Books 2007, p. 85

12.   Harry M Lawson, *Changes in Pesticide Use in the United
      Kingdom: Policies, Results and Long-Term Implications*, Weed
      Technology, Vol 8, No. 2 (Apr-June 1994) pp. 360–365

13. FERA, the Food and Environment Agency. Pesticide Usage Statistics, Pesticide Forum Indicator Report, published March 2008: http://pusstats.csl.gov.uk/index.cfm

14. EC Directive 76/464/EEC of 4th May 1976

15. www.environment-agency.gov.uk/business/topics/pollution/179.aspx

16. G Tyler Miller, *Living in the Environment,* Belmont: Wadsworth/Thomson Learning, 2002

17. Ritter SR. (2009). *Pinpointing Trends In Pesticide Use In 1939.*

18. www.opin.info/allpartybrief.php

19. Fergusson, *Ibid*, p. 63

20. Pesticide Action Network North America: www.panna.org/files/factsMalathion.dv.html

CHAPTER 5: ECOCIDE: THE 5TH CRIME AGAINST PEACE

1. Preamble, UN Charter, 1945.

2. The Rome Statute of the International Criminal Court entered into force on 1st July 2002. As of August 2015, 123 member states have ratified the Rome Statute, and a further 31 have signed but not ratified. A number of states including China, Russia and USA have not yet joined.

3. Article 5, Rome Statute.

4. The statute defines each of these crimes except for aggression, which has been the subject of discussion since the Rome Statute was first drafted. In June 2010 a definition was finally agreed. The Crime of Aggression is: '*committed by leaders who plan or execute an act of aggression that constitutes by its character, gravity and scale a manifest violation of the Charter of the United Nations.*' See Professor William A. Schabas' blog on the outcome of the recent ICC Review Conference: www.humanrightsdoctorate.blogspot.com/2010/06/kampala-review-conference-brief.html The outcome was a

compromise. See Amnesty International commentary: www.amnesty.org/en/news-and-updates/opt-out-system-risks-undermining-icc-2010-06-15

5.   The Longman Dictionary of Contemporary English defines ecocide as '*the gradual destruction of a large area of land, including all of the plants, animals etc living there, because of the effects of human activities such as cutting down trees, using pesticides etc [= ecological genocide]*'.

6.   See *Lessons Unlearned*, Global Witness Report 2010: an analysis of UN inability to prevent resource wars.

7.   Currently resulting in destruction, damage and loss of a territory the size of France.

8.   If proposed expansion proceeds, tar sand extraction will result in the loss of vast tracts of boreal forest and muskeg (peat bogs) of a territory the size of England.

9.   Globally, two million tons of sewage, industrial and agricultural waste is discharged into the world's waterways and at least 1.8 million children under five years-old die every year from water related disease, or one every 20 seconds. See: *Sick Water? A Rapid Response Assessment*. UNEP, UN-HABITAT report 2010.

10. *The Economics of Ecosystems and Biodiversity'* (TEEB), www.teebweb.org Financial analysis of the top corporations was provided by Trucost, www.trucost.com See also www.guardian.co.uk/environment/2010/may/21/biodiversity-un-report

11. Land deemed owned by no-one and therefore susceptible to acquisition.

12. Crimes of strict liability are crimes where the *mens rea*, the 'mental element', is irrelevant. Only the *actus reus,* 'the act of doing', is required. See Chapter 8 for further discussion on strict liability.

Chapter 6: The Sacred Trust of Civilization

1. Franklin Delano (ed.), *Nothing to Fear: The Selected Addresses of Franklin Delano Roosevelt*, Houghton Mifflin Company 1946, pp. 455–456

2. Article 73, Chapter XI: Declaration Regarding Non-Self Governing Territories, UN Charter, 1945.

3. E Burke, *Speech on Mr. Fox's East India Bill*, 1st December 1783

4. H. G. Prince, *The United States, the United Nations and Micronesia: Questions of Procedure, Substance and Faith*, 11 Michigan Journal of International Law (1989), p. 83.

5. See www.survivalinternational.org for their campaign against Vedanta

6. Although outside the scope of this book, it is of note that Chapters XI – XII have not been applied to territories in the aftermath of war despite their applicability. For an in-depth analysis see: Ralph Wilde, *Understanding the international territorial administration accountability deficit: trusteeship and the legitimacy of international organizations*, in Gian Luca Beruto (ed.), *International Humanitarian Law, Human Rights and Peace Operations*, International Institute of Humanitarian Law 2009, pp. 155–170.

7. Remaining on the list of Non-Self-Governing Territories are: Gibraltar, New Caledonia, Western Sahara, American Samoa, Anguilla, Bermuda, British Virgin Islands, Cayman Islands, Guam, Montserrat, Pitcairn, Saint Helena, Turks and Caicos Islands, United States Virgin Islands, Tokelau and the Falkland Islands (Malvinas).

8. Hans Kelson, *The Law of the United Nations, A Critical Analysis of its Fundamental Problems*, 1950. Kelson asserted that there was nothing to prevent the addition of new territories, p. 556.

Also referred to by Steven Hillebrink, *The Right to Self-Determination and Post-Colonial Governance, the Case of the Netherlands Antilles and Aruba,* T.M.C. Asser Press 2008, p. 6.

9. Kelson, *ibid* p. 554.

10. Kelson, *ibid* p. 554.

11. see Doc A/64. P13, ref Kelson, *ibid,* p. 554, footnotes 3 and 4.

12. Report of Commission I to Committee I: see UNCIO, Documents, vol.6, p. 455. See also Dr Sarah Pritchard's excellent article, *The Rights of Indigenous Peoples to Self-Determination under International Law,* Aboriginal Law Bulletin, [1992] Aboriginal LB 16, www.austlii.edu.au/au/journals/AboriginalLB/1992/16.html#fn10

13. UNCIO, Documents, Vol 18, p.657 f

14. Article 1, ICCPR, 1976, Article 1, ICESCR, 1976.

15. For analysis of self-determination by Kosovo see Ralph Wilde, *'Kosovo—Independence, Recognition and International Law,'* in *Kosovo: International Law and Recognition,* Chatham House 2008.

16. Dr Andrés Rigo-Sureda, *'The Evolution of the Right of Self-Determination. A Study of United Nations Practice',* Sijthoff 1973, p. 100

17. *'In practice the Security Council does not act on the understanding that its decisions outside chapter VII are binding on the States concerned. Indeed, as the wording of chapter VI clearly shows, non-binding recommendations are the general rule here':* Frowein, Jochen Abr. Völkerrecht - Menschenrechte - Verfassungsfragen Deutschlands und Europas, Springer 2004, p. 58

18. Advisory Opinion of the International Court of Justice on the Western Sahara. ICJ Reports (1975) 102, 105 ILR p. 12

19. ICJ Reports (1995) 122, 105 ILR p. 90

20. Debate as to what constituted colonialisation, a colony and a NSGT continued for many years within the UN, with no conclusive determination of definitions. See: Hillebrink, *ibid*, pp. 19–22.

21. Tilman Dedering, *Petitioning Geneva, Transnational Aspects of Protest and Resistance in South West Africa/Namibia after the First World War*, Journal of South African Studies, 35:4, pp. 785–801, provides an extensive analysis and documentation of the petitioning of the UN (and of it's precursor, the League of Nations) for remedy by the non-self-governing territory of the Reheboth Basters, who continually petitioned the UN from 1924 until 1961. Whilst the petitions did not provide immediate remedy, their petitioning could be called a success to some extent in two respects:
1. The peoples petitioned directly, without interference or refusal by the member state;
2. the UN imposed upon the member state the obligation to provide information, and failure to provide suitable reports within the timescales set down attracted international criticism, which in turn resulted in a staying of the unwarranted activity in the territory.
Crucially this instance serves as a precedent that petitioning by the people of a given territory in time of emergency will (and should) be responded to.

22. Palau, formerly part of the Trust Territory of the Pacific Islands.

23. Article 10, Charter of the UN, see also Kelson, *ibid*, p. 551.

24. Preamble, UN Charter, *ibid*.

CHAPTER 7: HOLDING BUSINESS TO ACCOUNT

1. Kenny Bruno, *The Corporate Capture of the Earth Summit*, www.multinationalmonitor.org/hyper/issues/1992/07/mm0792_07.html

2. For fuller discussion see Professor Philippe Sands, *Principles*

of International Environmental Law, Cambridge University Press, 2nd Edition, p. 620.

3. *R v Milford Haven Port Authority [2000]* 2 Cr App R(S) 423.

4. See Blackstone's *Commentary on the Laws of England*, 1765, p. 1455

5. www.onepercentfortheplanet.org

6. www.regjeringen.no/en/dep/bld/Topics/Equality/Rules-on-gender-representation-on-compan.html?id=416864

7. See www.csrgov.dk/sw51190.asp

8. See Catalyst for most up to date research on women in the workplace, www.catalyst.org

9. www.carbonmajors.com and see also: UN Women, 'Women in Politics: 2015' at: www.unwomen.org/en/what-we-do/leadership-and-political-participation/facts-and-figures

CHAPTER 8: ENVIRONMENTAL SUSTAINABILITY

1. *Cost Barriers to Environmental Justice*, Environmental Law Foundation 2010, www.eflaw.org

2. ACCC/C/2008/33 On 2nd December 2008, ClientEarth, the Marine Conservation Society (MCS) and Mr. Robert Latimer submitted a communication to the Committee, alleging non-compliance by the United Kingdom with its obligations under article 9, paragraphs 2, 3, 4 and 5 of the Aarhus Convention. Judgment was delivered on 24th August 2010. See in particular paras 128 and 143. www.unece.org/env/pp/compliance/C2008-33/DRF/C33DraftFindings.pdf

3. www.trust.org/trustlaw/pro-bono/news-and-analysis/detail.dot?id=5975e714-5b5e-46f1-bb42-57a5e3a0b4b5

4. William Hague, *William Wilberforce: The Life of the Great Anti-Slave Campaigner,* Harper Press 2007.

5. Trucost report, www.trucost.com www.guardian.co.uk/

environment/2010/may/21/biodiversity-un-report

6. Cain Burdeau and Holbrook Mohr, *BP Didn't Plan for Major Oil Spill*, ABC news 30th April 2010, www.abcnews.go.com/Business/wireStory?id=10515973

7. *Tools to Support Transparency in Local Governance*, Transparency International, p. 65 www.transparency.org/ tools/e_toolkit/tools_to_support_transparency_in_local_ governance

8. www.civicus.org/toolkits/legitimacy-transparency-and-accountability

9. www.glendaleaz.com

10. See Svetozar Pejovich, *The Economics of Property Rights: Towards a Theory of Comparative Systems*, Chapter 8, Kluwer Academic 1990, pp. 65–72

11. See World Resources Institute for facts and figures on species extinction, www.wri.org

12. Christopher Stone, *Should Trees Have Standing? Towards Legal Rights for Natural Objects*, Kaufmann Inc 1972, p. 11

13. Chris Goodhall, *Ten Technologies to Save the Planet*, Green Profile 2008, pp. 241–242

14. See Dr John Liu's documentary at www.earthshope.org

15. David Kennedy, *Over Here: The First World War and American Society*, Oxford University Press, 1980, p. 655

16. www.history.army.mil/documents/mobpam.htm

CHAPTER 9: NEW DEVELOPMENTS

1. Universal Declaration of Animal Welfare, www.animalsmatter.org

2. www.crimesagainstfuturegenerations.org.

3. World People's Conference on Climate Change: www.pwccc.wordpress.com

4. International Court for the Environment:
   www.environmentcourt.com

5. Annual Grotius Lecture to the British Institute of
   International and Comparative Law, Nov 2008. See also
   Lord Bingham's recent book, *The Rule of Law* and the
   BIICL Bingham Centre for the Rule of Law www.biicl.org/
   binghamcentre

6. www.pict-pcti.org

7. ICEF: www.icef-court.org

8. Resolutions B 30718/91 and B 302262/92

9. www.biopolitics.gr/HTML/PUBS/VOL8/html/Pirro.htm

10. Atmospheric Air Trust was developed by Professor Mary
    Wood, Faculty Director, Environmental and Natural
    Resources Law Program, University of Oregon School of Law.

11. www.globalcommonstrust.org, www.global-commons.net

12. Fred Harrison, *Ricardo's Law,* Shepheard-Walwyn 2006.
    Harrison makes the case for treating land values as a major
    source of public revenue. For a thorough analysis of the
    instability of our current economic system and how this can
    be rectified by a shift of taxes off enterprise onto land usage,
    see also Fred Harrison, *Boom Bust: House Prices, Banking
    and the Depression of 2010,* Shepheard-Walwyn 2007.

CHAPTER 10. THE COMMANDING VOICE OF THE PEOPLE

1. International Energy Agency World Energy Outlook 2009
   Fact Sheet www.iea.org/weo/docs/weo2009/fact_sheets_
   WEO_2009.pdf

2. BP's official oil sands resolution and response, March 2010,
   Market risks: Supply and Demand, p. 2
   http://tinyurl.com/yzl8z23

3. www.fairpensions.org.uk/press

Index

Probably the most important book in all history ...

Ecocide as a UN amendment could transform the face of the planet – as did the abolition of slavery in the C19.

I totally support and agree with Polly's goal for international legislation against ecocide and hope that winning this award will stir more public support.

This is an important, inspirational and visionary book which should be read by all world leaders!

Polly gets my vote. Our planet needs to be heard.

Enlightening and frightening – and essential reading. We need to think differently and faster.

This book needs to get out there.

Polly Higgins is an inspiration. An international lawyer, she has focused her talents towards the creation of a new international law, Ecocide ... This will be an incredibly difficult task but, having met her and realised the depth of her research and extent of her networking in Southern nations, if anyone can pull it off Higgins can.

This book should be compulsary reading for all people ... to inspire us all to lobby our leaders to ACT now and much much faster if necessary.

This is the most serious issue facing us; we must not deny it and Polly Higgins helps us face reality.

This book will help create a new foundation for a new humanity!

A book of universal influence and initiative that will stimulate the changes the planet needs.

Fabulous user-friendly legal book which manages to be easy and interesting to read, while being packed full of information. It's also a refreshing positive climate change book.

Making ecocide a crime is SO important that everything else pales into insignificance.

Best book this decade.

It's about time it was made a serious crime to cause ecocide.

An amazing idea! I hope this book leads to a healthier future for generations to come.

This is an important book and important legislation! The Earth is a Living and Intelligent Being, as we are. If we don't stop raping her and killing her, bit by bit; there will be no place for us to live.

I guess the survival of life on planet earth is a concept to get behind, eh?

A heroic and visionary book!

Maybe the most improtant book for the 21st century!

An inspiring read to shake us out of complacency.

This book proposes a method to bring to justice those who create wanton destruction to our environment ... Bravo!

A planet-saving read!

Polly Higgins pioneering visionary work is inspiring and this book is essential reading for us all to act on; this is just the start!

Polly Higgins is an activist lawyer and a true inspiration to all young lawyers.

Most needed book in these times ... Thanks Polly.

Essential reading. Ecocide must be eradicated or it will destroy us all.

The simple facts and the simple solution to guarantee a sustainable future for our children – a true call to action for every one of us.

Surely, this accessible yet comprehensive book counts among the most timely currently available on a crucial theme of our times.

Beyond important – we cannot carry on as if there is no tomorrow, as there will be no tomorrow.

This is a major contribution to high-efficacy strategy; to ignore it could be interpreted as a form of sui-genocide.

Polly knows law inside out. She also understands what matters in life. When these two come together, you have an explosive combination generating invaluable insights inspiring for action that counts.

Quite simply the best book on the most important topic of all time.

Polly Higgins is woman with a huge heart and great vision. She's waking up people all over the world to the need to protect our planet, and hence ourselves.

The simple fact of defining Ecocide as a crime could give pause to thinking members of [company] boards, and hold back destructive projects.

Truly inspiring – how laws could save the planet. This is insightful and proactive – it is as the forefront of a new wave of thought that offers real solutions to tackling global problems.